Food Webs and
Niche Space

MONOGRAPHS IN POPULATION BIOLOGY

EDITED BY ROBERT M. MAY

Food Webs and Niche Space

JOEL E. COHEN

PRINCETON, NEW JERSEY
PRINCETON UNIVERSTY PRESS

Contents

CONTENTS

CONTENTS

Preface

The aim of this work is to demonstrate that a niche space of dimension one suffices, unexpectedly often and perhaps always, to describe the trophic niche overlaps implied by real food webs in single habitats. Consequently, real food webs fall in a small subset of the set of mathematically possible food webs. That real food webs are compatible with one-dimensional niche spaces more often than can be explained by chance alone has not been noticed previously.

Other statistical features of real food webs have appeared in the course of this work. In the food webs studied here, which purport to describe a community, the ratio of the number of kinds of prey to the number of kinds of predators is close to a constant 3/4. The mean and variance of the number of overlaps among the predators' diets are reasonably described by assuming that every kind of predator has a constant and independent probability of preying on every kind of prey, with a probability that is characteristic of the food web.

This monograph does not aim to be a compendium of everything worth knowing about food webs and niche space.

This work began in the summer of 1968, when I went to the Mathematics Department of the RAND Corporation. I hoped to learn enough combinatorics to make sense of food webs. On 22 August 1968 I heard a lecture by my college classmate Alan C. Tucker, which included the definition of interval graphs. The idea of using interval graphs to study the dimensionality of ecological niche space appeared to me in a dream the next morning. In the next week I applied the idea to several

PREFACE

real food webs and wrote up a general conjecture in an internally circulated document, "Interval graphs and food webs: a finding and a problem," RAND Document 17696-PR, 30 August 1968. Klee (1969, p. 811) published the conjecture. Roberts (1976c; in press) repeats the conjecture, perhaps more apodictically than the evidence now warrants. I first offered evidence favoring a qualified form of that conjecture in 1977. I no longer claim that niche overlap inferred from a food web is a priori evidence for competition (section 2.4).

My only regret about taking so many years to do this work is that I cannot now ask for the criticisms of D. Ray Fulkerson, who died on 10 January 1976 at the age of 51. His kind guidance at Rand in 1968 was enormously encouraging.

For detailed comments on the manuscript and many useful suggestions and references, I am indebted to Victor Klee, Thomas Mueller, Robert T. Paine, and Thomas W. Schoener. For helpful conversations and correspondence, before and after I drafted the manuscript, I am grateful to Mike Apl, Robert Donaghey, W. T. Edmondson, Joseph Felsenstein, D. Ray Fulkerson, Hyman Gabai, Michael Gilpin, Nelson G. Hairston, Joel W. Hedgepeth, Alan J. Kohn, Janos Komlos, Richard C. Lewontin, Robert M. May, Jane Menge, Joel S. O'Connor, Gordon H. Orians, H. Ronald Pulliam, John Riordan, Fred S. Roberts, L. B. Slobodkin, H. P. Swinnerton-Dyer, Robert L. Trivers, Alan C. Tucker, Leigh Van Valen, George C. Williams, and Edward O. Wilson. For help in collecting food webs, I thank Paula Pinkston and Chaiya Wongkrajong. Thomas Mueller assisted in programming and computing; he implemented the algorithm of Fulkerson and Gross (1965) in APL (Gilman and Rose, 1974). Anne Whittaker prepared the typescript and helped editorially. Gail Filion edited wisely.

Without the freedom to pursue research of The Rockefeller University, this work would have remained a morning's dream.

I dedicate this book to Zoe Sara.

List of Tables

List of Figures

Food Webs and
Niche Space

CHAPTER ONE

Introduction

> From the least to the greatest in the zoological progression, the stomach sways the world; the data supplied by food are chief among all the documents of life.
>
> J. H. Fabré, 1913 (from Elton 1966, p. 38)

Ecological studies of where the organisms in communities are and what the organisms do, especially what they eat, frequently use the concept of an ecological niche. Hutchinson (1965, p. 27) traces the concept back to Steere and Grinnell at the beginning of this century. Studies of what organisms eat frequently also use the concept of a food web. Early food web graphs, based unfortunately on hypothesized communities, appear in Shelford (1913).

Both concepts, of ecological niche and of food web, have been elaborated subsequently (R. S. Miller, 1967; Vandermeer, 1972; Pianka, 1976).

Hutchinson (1944; quoted by Miller, 1967, p. 16) defines the ecological niche as "the sum of all the environmental factors acting on an organism; the niche thus defined is a region of n-dimensional hyper-space, comparable to the phase-space of statistical mechanics." By "niche space" we mean just this n-dimensional hyperspace; Hutchinson (1965) gives numerous examples and applications.

By a food web, we mean a set of different kinds of organisms, together with a relation that shows the kinds of organisms, if any, that each kind of organism in the set eats. A whole range of sophisticated questions will be deferred temporarily. (These questions include: What is a "kind" of organism? How does one decide which "kinds" to include in a set called a community? Does "eat" mean all the time, or only sometime? Does "kind A eats kind B" mean all members of A must

3

eat members of B, or only some?) Without necessarily consistent or explicit answers to these questions, ecologists have reported many food webs. We accept these reports as evidence that operational definitions of the concept of food web exist. Gallopin's (1972, p. 245) "eating relation" formalizes the concept of food web in a way that suits our purpose. We restate the essence of that formalization in section 2.1.

This essay has four purposes. One is to present a technique for obtaining a partial answer to an elementary question about niche space: What is the minimum dimensionality of a niche space necessary to represent the overlaps among observed niches? "Represent" is defined in Chapter 2.

The second purpose is to propose an empirical generalization, based on the application of this technique to data, about both food webs and niche space. The empirical generalization proposed is: Within habitats of a certain limited physical and temporal heterogeneity, the overlaps among niches along their trophic (feeding) dimensions can be represented in a one-dimensional space far more often than expected by chance alone. To support this generalization, it was necessary to assemble observed food webs from ecological publications in a consistent, machine-readable form.

A third purpose is to exploit the collected food webs to test old empirical generalizations about food webs and to present some new ones.

A fourth purpose is to make plain the limitations of this approach. Some procedures and forms of reporting are suggested to field ecologists to put the ideas here to a sharper test.

Chapter 2 presents a technique for relating the combinatorial structure of food webs to the minimal dimension of a space needed to represent the overlap along trophic dimensions of ecological niches, and illustrates the technique with examples.

In Chapter 3, this technique is applied to real food webs.

In Chapter 4, statistical features of these food webs are studied in order to estimate the universe of food webs from which the observed food webs may be drawn.

Chapter 5 investigates quantitatively whether one ought to be surprised by the results of Chapter 3, by estimating the proportion of interval graphs in several universes of possible graphs.

Chapter 6 offers physical interpretations of an inferred one dimension in the niche space of communities in single habitats. It discusses some theoretical consequences of and rationales for a one-dimensional niche space.

Chapter 7 suggests extensions of the present approach and criticizes the approach and results.

Chapter 8 collects concrete suggestions for future studies.

CHAPTER TWO

A Relation between
Food Webs and Niche Space

I passed by his garden, and marked, with one eye,
How the Owl and the Panther were sharing a pie:
The Panther took pie-crust, and gravy and meat,
While the Owl had the dish as its share of the treat.
When the pie was all finished, the Owl, as a boon,
Was kindly permitted to pocket the spoon:
While the Panther received knife and fork with growl,
And concluded the banquet by—
                    Lewis Carroll, *Alice in Wonderland*

2.1 INTERVAL GRAPHS AND THE DIMENSIONALITY OF NICHE SPACE

If two kinds of organisms, say A and B, both eat some kind of organism C (not necessarily distinct from A or B, in the case of cannibalism), then along some trophic dimensions, the niches of A and B logically must overlap or intersect. The niches of A and B need not overlap along physical dimensions of niche space. For example, if an insect C that lives on the surface of fresh water is eaten by an aquatic fish A and by a bird B, it does not follow that the ranges of oxygen tension in which the bird and the fish can survive overlap. For this reason, "niche overlap" here means "niche overlap along the trophic dimensions" unless explicitly stated otherwise. "Dimensions" is plural here because several factors may appear to influence a predator's diet.

The distinction between the trophic and the other dimensions of niche space may be viewed as a special case of Whittaker's (1969, pp. 184–185) distinction between "niche factors" and "habitat factors."

Our operational definition of the way organisms in a community are allocated into "kinds" is the method of allocation

6

practiced by ecologists who reported the food webs we use as data. From examination of these food webs, we infer that a condition that often suffices for ecologists to consider a group of organisms as a single kind in a food web is that all organisms in the group have identical kinds of organisms as predators and identical kinds of organisms as prey. "Kinds" are equivalence classes with respect to trophic relations.

We assume that a group of organisms qualifies as one "kind" of organism in a food web only if its niche, viewed as a region or set of points in niche space, is connected (along the trophic dimensions), that is, only if it is possible to pass from any one point in the niche to any other without leaving the niche. (We treat the niche as a region in niche space, not necessarily as a directly observable region of space and time.) For example, if two stages in the life cycle of a single species of insect were so different that the region in niche space corresponding to one stage were unconnected to the region corresponding to another, we assume that the two stages would have feeding habits sufficiently different that the stages would be distinguished as different "kinds" in a food web. Hardy's (1924) study of herring makes this distinction.

Root's (1975) "guild" describes a grouping that we believe to be broader than our "kind" of organism.

In a one-dimensional space, which can be thought of as a straight line, the only connected regions are points and intervals. Since we care only about the overlaps among niches, we can always replace points by intervals. Consequently, if the niche overlaps of a set of kinds of organisms can be represented in a one-dimensional space, and if the niche of any one kind of organism is a connected region, then relations of niche overlap can be represented by the relations of overlap among intervals on the real line. Conversely, suppose there is a set of intervals on the real line, each one carrying the label of one kind of organism. Suppose that two intervals A and B overlap if and only if the niches of A and B overlap (once again, along the trophic dimensions). Then the niche overlap can be represented in a one-dimensional space, or is consistent with a one-dimensional niche space.

To make this argument concrete, we now analyze two food webs, one based on real observations and one hypothetical.

Figure 1 is a redrawing of Figure 9 of Bird (1930, p. 393), "a diagram of biotic interaction in the willow communities." Certain redundancies in Bird's figure have been eliminated, but the placement of each "kind" of organism (those repre-

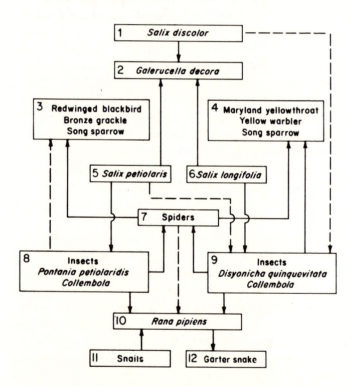

FIGURE 1. The food web graph of the willow forest, Canada. The arrows are directed from the kind of organism eaten to the kind of organism that eats. The dashed arrow are tentative. (Adapted from Bird, 1930, p. 393.)

sented in a single numbered box) has been retained approximately to facilitate comparison of this Figure 1 with the original. The appearance of the song sparrow in box [3] and box [4] must be interpreted to mean that it alone of the five bird species in those boxes eats both *Pontania petiolaridis* and *Disyonicha quinquevitata* (Thomas W. Schoener, personal communication, 1977). Presumably the plural form of snails appears in box [11] to indicate several species.

We follow Bird's convention (p. 393) that "arrows point from the animals or plants eaten to the animals that eat them," rather than the opposite convention adopted by Gallopin (1972). This is the only respect in which our graph formalization of the concept of food web differs from Gallopin's.

The solid arrows in Figure 1.1 include all the arrows from Bird's figure plus some feeding relations that can be unambiguously inferred from Bird's text but that were omitted from his figure. The dashed arrows are feeding relations that could be tentatively inferred from other food webs in the same article or from plausible interpretations of ambiguous passages in Bird's text.

In the analysis of this and the other food webs studied in Chapter 3, we do not claim firsthand knowledge of the natural history. Our purpose is not to revise or correct the observations of these earlier authors but to use them after a careful reading. We hope to make our analysis clear enough that others who find these food webs in error can carry out their own revised analysis.

In Figure 1, in the top left corner of each box is a number that has been assigned to the "kind" of organism in that box for convenient reference in the following figures. From Figure 1 we construct the niche overlap graph in Figure 2. There is a line connecting two kinds of organisms in Figure 2 if and only if both kinds eat some kind of organism in common in Figure 1. Such a line, without an arrow or direction, is known in the terminology of graph theory as an undirected edge. For ex-

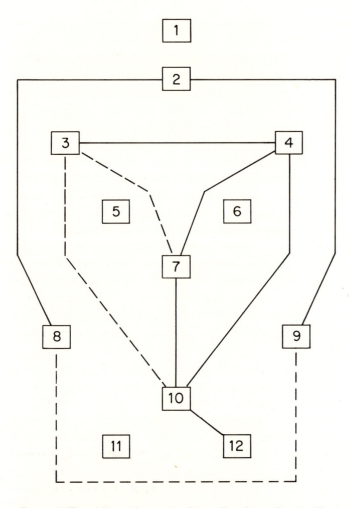

FIGURE 2. The niche overlap graph of the willow forest, Canada. Two kinds of organisms are joined by an undirected edge if their diets in Figure 1 overlap. The dashed edges are overlaps inferred only from tentative information. (Based on Figure 1.)

ample, *Salix discolor* eats nothing in Figure 1, so it has no niche overlap with any other kind of organism in Figure 2. The Maryland yellowthroat, yellow warbler, and song sparrow [4] eat insects, *Disyonicha quinquevitata*, and *Collembola* [9] and so do spiders [7] in Figure 1. Since the eating arrows are both solid lines in Figure 1, the overlap between [4] and [7] in Figure 2 is drawn with a solid edge. When the overlap between two niches is established only by a pair of eating arrows at least one of which is dashed, then the undirected edge between the two kinds in Figure 2 is also dashed.

Can all the overlaps of niches in Figure 2 be represented exactly by connected regions in a one-dimensional space, or simply by intervals on a line? Figure 3 shows that the answer

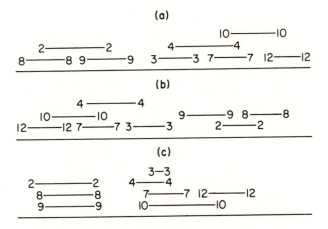

FIGURE 3. (a) An interval representation of the solid-edge niche overlap graph in Figure 2. Two intervals overlap here if and only if the corresponding kinds of organisms are joined by a solid edge in Figure 2. (b) Another interval representation of the identical graph. (c) An interval representation of the solid- and dashed-edge niche overlap graph in Figure 2. Two intervals overlap here if and only if the corresponding kinds of organisms are joined by a solid or dashed edge in Figure 2. In this example, the intervals of herbivores are separable from those of carnivores. (Based on Figure 2.)

is: Yes. Figure 3a shows one of many possible arrangements of intervals that represent the solid-edge overlaps in Figure 2. The intervals are shown sitting above the real line to avoid confusion, but are to be thought of as segments of the line. Each interval is labeled at both ends by the kind of organism to which it corresponds.

Figure 3a represents the solid overlaps in Figure 2 in this sense: in Figure 3, interval 2 overlaps intervals 8 and 9 but no others; in Figure 2, niche [2] overlaps niches [8] and [9] but no others. To emphasize that the ordering of intervals on the line is not uniquely determined by the overlaps among them, Figure 3b represents exactly the same overlaps as Figure 3a in another sequence. Figure 3c shows that all the niche overlaps, whether solid or dashed, can be represented by intervals of a line. Niches [8] and [9] overlap in Figure 3c but not in Figures 3a,b.

By a graph we mean a set of points (which might be drawn as the numbered boxes in Figures 1 and 2) and a set of edges between some, all, or none of them. If each edge has an arrow indicating a direction, then the graph is called a directed graph. The directed graph in Figure 1 is a *food web graph*. If no edge has an arrow indicating a direction, then we call the graph just a graph, or sometimes an *undirected graph*, for clarity. We call the graph in Figure 2 the *niche overlap graph* of the food web graph in Figure 1.

A graph is called an *interval graph* if and only if to each point in the graph there corresponds an interval of the real line, and two points in the graph are joined by an edge when and only when the two corresponding intervals overlap. Such a graph is also known as the *intersection graph* of the set of intervals (section 7.1.1). The edges of the graph represent the overlaps among the intervals, and vice versa.

Thus the food web graph of the willow communities reported by Bird has a niche overlap graph that is an interval graph, whether one uses only the solid eating arrows or both the solid and dashed arrows. This means that the relations of overlap can be represented in some one, as yet unidentified dimension.

Figures 1 to 3 are equivalent to other symbols that are much more convenient for machine computation. Corresponding to the food web graph in Figure 1 is the food web matrix in Table 1. The columns are labeled by the corresponding kind of eating organism or predator or parasite, and the rows are labeled by the corresponding kind of organism eaten or prey or host. There is a 1 at the intersection of a row and column where and only where there is a solid arrow from the prey to the predator in Figure 1. There is a −1 for each dashed arrow. Rows and columns containing only zeros have been omitted.

TABLE 1. The food web matrix of the willow forest, Canada, according to Bird (1930, p. 393).

Prey	Predators							
	2^a	3	4	7	8	9	10	12
1	1^b	0^c	0	0	0	-1^d	0	0
5	1	0	0	0	1	−1	0	0
6	1	0	0	0	0	1	0	0
7	0	1	1	0	0	0	−1	0
8	0	−1	0	1	0	0	1	0
9	0	0	1	1	0	0	1	0
10	0	0	0	0	0	0	1	1
11	0	0	0	0	0	0	1	0

a The key that identifies the kind of organism referred to by this number is under food web 1.2 in Appendix 1.

b The predator in this column eats the prey in this row.

c The predator in this column does not eat the prey in this row.

d It is tentatively inferred that the predator in this column eats the prey in this row. In food web version 1.21, the entry −1 is replaced by 0; in version 1.22, by 1.

Ecologists have long represented feeding relations in forms like Table 1. Kohn (1959) gives quantitative frequencies of predation in such tables. His tables may be reduced to the form of Table 1 by replacing each of his positive numbers with 1 and leaving each of his zeros as 0. The result of replacing each non-zero element in a matrix by 1 is called the incidence matrix of the original matrix. Clearly Figure 1 suffices to construct Table 1 and vice versa.

Because the food web graph and the food web matrix describe the same eating relation, we refer to that relation as the food web without specifying whether it is represented by a directed graph or a matrix of 0's and 1's.

Table 2 presents the niche overlap matrix constructed from Table 1. In Table 2 both the rows and columns correspond to predators. There is a 1 at the intersection of row [3] and column [4] in Table 2 because there is a row, namely row [7], in Table 1 that has a 1 in both columns [3] and [4]. There is a 0 in row [4] and column [8] of Table 2 because there is no row in Table 1 that has a 1 in both columns [4] and [8]. The additional overlaps between columns in Table 1 that would

TABLE 2. The niche overlap matrix of the willow forest, Canada, based on Table 1.

Predators	2^a	3	4	7	8	9	10	12
2	1	0^b	0	0	1^c	1	0	0
3	0	1	1	-1^d	0	0	−1	0
4	0	1	1	1	0	0	1	0
7	0	−1	1	1	0	0	1	0
8	1	0	0	0	1	−1	0	0
9	1	0	0	0	−1	1	0	0
10	0	−1	1	1	0	0	1	1
12	0	0	0	0	0	0	1	1

[a] The key that identifies the kind of organism referred to by this number is under food web 1.2 in Appendix 1.

[b] The predator in this column takes no kind of prey organism in common with the predator in this row, according to both food web versions 1.21 and 1.22.

[c] The predator in this column takes at least one kind of prey organism in common with the predator in this row, according to food web version 1.21.

[d] The predator in this column takes at least one kind of prey organism in common with the predator in this row, according to food web version 1.22 but not according to version 1.21.

14

arise if every −1 there were replaced by 1 are recorded as −1 in Table 2. Figure 2 suffices to construct Table 2 and vice versa.

By a clique, we mean a set of points in a graph such that there is an edge between every pair in the set. For example, points [4], [7], and [10] in Figure 2 form a clique. A dominant clique is a clique that is not contained in any larger clique. For example, if we recognize the dashed as well as the solid edges in Figure 2, then [4], [7], and [10] are not a dominant clique because they are a proper subset of [3], [4], [7], and [10], which form a dominant clique.

Table 3a lists the dominant cliques calculated from Table 2 with −1 replaced by 0. Table 3b lists the dominant cliques

TABLE 3. The dominant clique matrix of the willow forest, Canada, based on Table 2.

3a

		Predators						
	2[a]	3	4	7	8	9	10	12
	1[b]	0	0	0	1	0	0	0
Dominant	1	0	0	0	0	1	0	0
cliques	0	1	1	0	0	0	0	0
	0	0	1	1	0	0	1	0
	0	0	0	0	0	0	1	1

3b

		Predators						
	2	3	4	7	8	9	10	12
Dominant	1[c]	0	0	0	1	1	0	0
cliques	0	1	1	1	0	0	1	0
	0	0	0	0	0	0	1	1

[a] The key that identifies the kind of organism referred to by this number is under food web 1.2 in Appendix 1.

[b] According to food web version 1.21, the diet of the predator in this column overlaps the diet of every other predator for which there is a 1 in this row. The set of predators specified by the entries of 1 in this row is not contained in any larger set of predators all pairs of which have overlapping diets.

[c] Same as b, according to food web version 1.22.

15

calculated with each -1 replaced by 1. In Table 3 the columns correspond to columns in Table 2, i.e., to predators. Each row in Table 3 lists the membership of a dominant clique, with a 1 where the predator is included and a 0 where the predator is excluded. All the information about niche overlap in Table 2 is contained in the so-called dominant clique matrix of Table 3, and vice versa. Hence the dominant clique matrix completely describes the overlaps in Table 2 and Figures 2 and 3.

Each dominant clique in Table 3 might correspond to a trophic level, a concept as ambiguous as the concepts of food webs, niches, and communities. For example, the first line of Table 3b is the dominant clique containing all phytophagous insects. The second contains insectivores, and the third carnivores. This result is not surprising, since every pair of predators in a clique has at least one overlapping item of diet.

If one chooses any column of Table 3a or 3b and scans the entries from top to bottom, one never finds two 1's separated by a 0. Hence this matrix has what is called the consecutive 1's property. A matrix of 0's and 1's whose rows can be permuted so that in any column no two 1's are separated by one or more 0's is said to have the consecutive 1's property. Fulkerson and Gross (1965) proved that a graph is an interval graph if and only if its dominant clique matrix has the consecutive 1's property.

The intuitive foundation of the theorem of Fulkerson and Gross (1965) and the correspondence between Table 3 and Figure 3 are clear if the reader runs a pencil line down the 1's in each column of Table 3, skipping the 0's. If Table 3 is then rotated 90° to the left, the overlaps among the pencil lines correspond exactly to the overlaps among the corresponding intervals in Figure 3.

All the results about interval graphs reported here were obtained by a machine implementation (in APL, executed on the Cornell University Medical College IBM 370/145) of the algorithm of Fulkerson and Gross (1965) for determining whether a matrix has the consecutive 1's property. The results were checked by hand computation in several hundred cases. If n is the number of vertices in a graph or the number of columns in a dominant clique matrix (here n is the number of kinds of predators in a food web), then in the worst cases the number of steps the algorithm of Fulkerson and Gross (1965) requires to determine whether the graph is interval is proportional to at least n^3 as n gets large. Using this algorithm, the expense of computation increases rapidly with the number of kinds of predators. Booth and Lueker (1976; Booth, 1975) have discovered an algorithm for testing whether a graph is interval in which the time of computation increases in the worst cases at a rate proportional to $n + E$, where E is the number of (undirected) edges in the graph. A recent implementation of their algorithm appears to test economically graphs with at least 150 vertices.

Not every conceivable food web implies niche overlaps that can be represented in one dimension. Lekkerkerker and Boland (1962) and Gilmore and Hoffman (1964) have shown which subgraphs are forbidden in an interval graph.

To give the simplest example of a food web with a non-interval niche overlap graph, Figure 4 presents a hypothetical food web containing four species of rusts [1] to [4] and four species of trees [5] to [8]. The niche overlap graph (Figure 5) based on Figure 4 is not an interval graph. Choose any interval for rust [1]. Choose any interval overlapping the first and extending to the right for rust [2]. Then the interval for rust [3] must overlap that for rust [2] but not that for rust [1], and hence must be disjoint and to the right of interval 1. Now the interval for rust [4] must overlap the interval for rust [3] and the interval for rust [1] without overlapping the interval for rust [2], which is clearly impossible. W. C. Denison of the Department of Botany, Oregon State University (personal communication, 1968) constructed the food web in Figure 4 using a compendium of hosts and parasites. However, at least in his

17

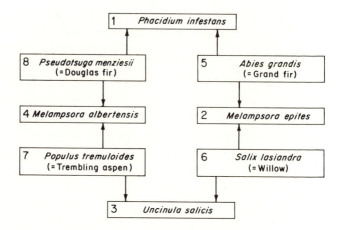

FIGURE 4. A hypothetical food web graph of four species of rusts [1] to [4] and four species of trees [5] to [8]. (Adapted from W. C. Denison, personal communication, 1968.)

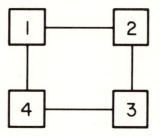

FIGURE 5. A hypothetical niche overlap graph that is not an interval graph. (Based on Figure 4.)

example, the four trees are not found together in any natural "community." At least some of the rusts are obligate parasites on the two different trees in their diet at different stages in their life cycles. Hence Denison's construction would not be observed at any single point in time in any one place.

The niche overlap graph in Figure 5 can be represented by the overlaps of two-dimensional boxes (Figure 6). We shall return to higher dimensional extensions in section 7.1.1.

18

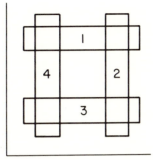

FIGURE 6. A representation of the hypothetical niche overlap graph in
Figure 5 by 2-dimensional boxes. Two boxes overlap here if and only if
the corresponding rusts in Figure 4 parasitize a tree in common. (Based
on Figure 5.)

Interval graphs have already helped three other life sciences
in evaluating the dimensionality of structures known only from
information about the overlaps of components. First, Benzer
(1959) uses this approach to determine whether genetic fine
structure could be linear (one-dimensional). Second, in psychology, Roberts (1969b) and Hubert (1974) relate this approach to determining the dimensionality of perceptions in
psychophysical experiments, the dimensionality of preferences
in economics, and the sequencing of subjects using dichotomous attributes. Third, Kendall (1969) explores the use of
interval graphs in determining an ordering in time of archaeo-
logical artifacts found in various strata. Booth (1975) cites
other equally diverse applications of interval graphs.

In the previous and present applications, the finding that a
graph is an interval graph does not prove the one-dimensionality of the structure (here niche space) whose components
correspond to the points (here niches) of the graph. Finding an
interval graph only proves that the overlaps among the components could be represented in a one-dimensional space.
In the applications to genetic fine structure and archaeology, physical distance along a molecule and time are natural

physical interpretations for the inferred single dimension. In the psychological, economic, and present ecological applications, an inferred single dimension may correspond to some measurable property, such as the wavelength of colors, the price of goods, or the physical size of prey; or further investigation may be required to identify measurements that correspond to an inferred dimension.

The finding that a graph is not an interval graph *does* prove that the structure whose components correspond to the points of the graph cannot be one-dimensional, assuming that the graph represents the overlaps of the components without error. (We return to the problem of errors of observation in sections 7.1.2 and 7.2.2.) The technique of testing whether a graph is an interval graph is thus a strong technique for rejecting the hypothesis that a certain structure is one-dimensional.

To avoid future circumlocutions, we shall say that a food web is an interval food web, or that a food web is interval, if its niche overlap graph is an interval graph. This is an abuse of language because the food web graph is a directed graph, and only the undirected niche overlap graph can have (or fail to have) the property of being representable by intervals. Nevertheless the abbreviation is convenient. A food web will be called a non-interval food web, or non-interval, if its niche overlap graph is not an interval graph.

2.2 FOOD WEBS DEFINED BY COMMUNITY, BY SOURCES, AND BY SINKS

Ecologists select the kinds of organisms that appear in food webs in three ways that differ significantly for our purposes. We label the food webs resulting from these approaches "community" food webs, "sink" food webs, or "source" food webs.

A community food web is defined by picking, within a habitat or set of habitats, a set of kinds of organisms. The set of kinds of organisms is chosen, without prior regard to the eat-

20

ing relations among them, by taxonomy, size, location, or other criteria not explicitly dependent on eating relations. This set of kinds of organisms is interpreted as the set of vertices of the community food web. All the eating relations among those kinds of organisms are taken as the directed edges of the community food web.

This definition of a community food web requires an interpretation of the concept of habitat. "By definition a habitat must possess uniformity with respect to an important quality" (Kohn, 1959, p. 63). In some cases a habitat may be physically defined, as by a pond, island, edaphic conditions, or a range of temperatures in a hot spring. In other cases a habitat may be defined biologically in terms of characteristic organisms. For example, the habitat might be defined to include as its points the animals living in a forest characterized by certain kinds of trees.

A sink food web is a directed subgraph of a community food web. It includes one or more kinds of organisms (who are called the "consumers") plus all the kinds of organisms that the consumers eat. The prey of the prey of the consumers, and so on, are also among the kinds of organisms in a sink food web. The directed arrows in a sink food web are all the directed arrows in the community food web that go from one point to another in the sink food web. A sink food web is so named because it is defined by the sinks for energy or biomass flow.

A community food web is interval if and only if every possible sink food web contained in it is interval. To prove this theorem, suppose first that every possible sink food web contained in a community food web is interval. Then the sink food web containing all the points of the community food web, which is identical to the community food web, is interval.

On the other hand, suppose that a community food web is interval. Consider the edges in the niche overlap graph of a sink food web contained in this community food web. These edges describe all the niche overlaps of the organisms in the

21

sink food web based on all the food those organisms eat. Hence if one added to the sink food web additional kinds of organisms from the larger community food web, one would not change the niche overlaps among the kinds of organisms originally in the initial sink food web. Thus one could continue to add kinds of organisms to the initial sink food web until one had the entire community food web, without altering the initial niche overlaps of the kinds of organisms in the initial sink food web. Now if the niche overlap graph of the initial sink food web were not an interval graph, it would remain impossible to represent by intervals the overlaps among those kinds of organisms in the community food web, and hence the niche overlap graph of the community food web would not be an interval graph, contrary to assumption. So every sink food web must be interval.

The practical impact of this theorem is that, if even one sink food web is non-interval, then the community food web it came from is non-interval.

A source food web is also a directed subgraph of a community food web. It includes one or more kinds of organisms (not necessarily primary producers in the usual sense, but the sources of energy or biomass for that subweb), plus all the kinds of organisms that eat those sources, plus all the predators of those predators, and so on. The directed arrows in a source food web are all the directed arrows in its community food web that go from one point to another in the source food web.

A community food web may be interval at the same time a source food web contained in it is non-interval. To give a hypothetical example, suppose a community food web consists of the food web in Figure 4 plus another tree [9] that is fed upon only by rust [1] and rust [3]. The niche overlap graph would be as shown in Figure 5 with an additional edge between [1] and [3]; such a graph would be an interval graph. However, the source food web based on the first four trees, namely, [5] to [8], would have just the niche overlap graph shown in Figure 5, which is not an interval graph.

22

Similarly, a community food web may not be interval at the same time that a source food web in it is interval. For example, suppose the community food web is given by Figure 4 with niche overlap graph given by Figure 5. The source food web based on [6], [7], and [8] has the niche overlap graph obtained by removing the edge between [1] and [2] from Figure 5, which then becomes an interval graph.

The difference between community food webs, on the one hand, and source and sink food webs, on the other, is that, in a community food web, all the vertices (kinds of organisms) are chosen before observing the eating relations among them. In source (or sink) food webs, a group of sources (or sinks) is chosen first, and other kinds of organisms are included as vertices in the food web depending on their eating relations, through one or several directed edges, with the initial set of sources (or sinks).

The usefulness of a food web for testing a hypothesis depends strongly on how the food web is constructed. In particular, source food webs are uninformative about the minimal dimensionality needed to represent niche overlap.

Hence, in choosing food webs for empirical study here, those arising in agricultural ecology from studies of the consumers of a crop of interest were omitted (e.g. Root's [1975] careful studies of the consumers of collard foliage). Food webs describing the diet of a kind or set of kinds of organisms were sought.

2.3 FOOD WEBS OF SINGLE HABITATS AND OF COMPOSITE COMMUNITIES

In advance of evidence, suppose that every niche space within a habitat, as defined above by Kohn (1959), is one-dimensional.

It is possible, but less likely than an alternative that we will describe, that in several distinguishable, geographically juxtaposed habitats the single dimension of each niche space is the same for all habitats and is also the same as the feature that distinguishes one habitat from another. For example, it is possible that in a sandy beach habitat, a surface water habitat, and a deep water habitat, the single dimension of niche space in all three habitats is temperature, and that temperature also distinguishes one habitat from another.

It seems a priori more likely, however, that the dimension of niche space in one habitat will not be identical to the dimension in another: temperature might be the dimension in one habitat, light intensity in another. It seems even more likely that the features differentiating one habitat from another will be multidimensional (Cody, 1968; Schoener, 1974) and different from the dimension of variation within a habitat.

Consequently, one should not necessarily expect food webs of composite communities, that is, communities containing more than one habitat, to be interval even if niche space within any given habitat were one-dimensional.

2.4 COMPETITION AND NICHE OVERLAP

Many notable authors have proposed definitions and interpretations of competition in ecology. We consider the relation of some of these to the niche overlap that food webs reveal.

Hartley (1948, p. 4) proposes a relaxed definition: "In the absence of data upon the density of the food organisms, and the proportion removed by fish predation in the course of the year, 'competition' here means nothing more exact than 'taking an appreciable proportion of a common food', without any implication that the supply of that food was a limiting factor." Assuming that "appreciable" means "large enough to be observed," competition in Hartley's sense is identical to niche overlap.

Miller (1967, p. 6) quotes the definition of Clements and Shelford (1939): interspecific competition "is the active de-

24

mand by . . . members of 2 or more species at the same trophic level . . . for a common resource or requirement that is
actually or potentially limiting." This definition rests too
heavily on the ambiguous concept of "trophic level" and the
weasel word "potentially" to be helpful. But it does suggest
that competition could occur for some resource other than
food, and therefore that (trophic) niche overlap need not be
identical to competition.

May (1975) clarifies the relations among several formulas
proposed to estimate competition. These formulas are based
on measures of the relative use of resources by each population in a community. When the resource categories can be
classified into distinct dimensions, these dimensions can be
identified as dimensions of the niches of the populations (May,
1975, pp. 738–39). According to these formulas, competition
between two populations is positive if at least one resource
category is utilized by both populations. Competition is zero
otherwise. Therefore if the resource categories pertain to food,
but are not necessarily chosen as "limiting" resources, these
formulas give quantitative form to Hartley's definition and
to our concept of (trophic) niche overlap, which rests on a
dot product of prey utilization. If the resource categories do
not all pertain to food, and if the categories are chosen as
"actually or potentially limiting," then these formulas embody
the definition of competition of Clements and Shelford (1939),
and need not be identical to our concept of niche overlap.

MacArthur's (1972, p. 21) "wide definition of competition:
two species are competing if an increase in either one harms
the other," leaves competition undefined for sets of organisms
other than species. Since it defines competition in terms of an
increment in one species, the definition makes it impossible to
say that two species are competing without a natural or artificial perturbation experiment. Worse (Thomas W. Schoener,
personal communication, 1977), if species A eats species B and
species C, and if an increase in B leads to an increase in A which

leads to a decrease in C because C is relatively poor at avoiding predation by A, then B and C would be defined as competing by this definition.

We prefer Pianka's (1976, p. 114) simple definition and an apparently standard distinction he makes between two kinds of competition: "competition occurs when two or more organisms, or other organismic units such as populations, interfere with or inhibit one another. . . . Competition is sometimes quite direct, as in the case of interspecific territoriality, and is then termed *interference competition*. More indirect competition also occurs, such as that arising through the joint use of the same limited resources, which is termed *exploitation competition*." When one of the limited resources is food, niche overlap along the trophic dimensions occurs. We accept a special case of Pianka's conclusion (1976, p. 122): "niche overlap in itself is neither a necessary nor a sufficient condition for interference competition; moreover, [trophic niche] overlap is only a necessary but not a sufficient condition for exploitation competition" when food is limiting and other resources are not limiting.

Therefore a low level of exploitation competition may be inferred when a low level of niche overlap is observed in food webs; but a high level of niche overlap implies only the possibility of a high level of exploitation competition. Empirically, a low level of niche overlap may be an evolutionary consequence of a high level of interference competition even in the absence of a logically necessary relation (Thomas W. Schoener, personal communication, 1977).

2.5 SUMMARY

The niches of two kinds of organisms must overlap, at least along the trophic dimensions of their ecological niche space, if there is some kind of organism that they both eat. Hence the overlaps (along trophic dimensions) among ecological niches can be inferred from food webs and described by an undirected "niche overlap graph." If the niche of a kind of organism is a connected region in niche space, then it is possible for niche overlaps to be described in a one-dimensional niche space if and only if the niche overlap graph is an interval graph. We give worked examples.

Food webs may be characterized as describing a single habitat or as describing a composite of several habitats. Food webs describing a composite of habitats are not expected to be interval even if the niche overlap graphs of the component individual habitats can be represented in one dimension.

Single-habitat food webs may also be characterized as attempting to describe all the kinds of organisms (possibly restricted to some location, size, or taxa) in a habitat ("community food webs"); or all the prey taken by a set of predators ("sink food webs"); or all the predators on a set of food sources ("source food webs"). Source food webs are not informative about the dimensionality of niche space; the other two kinds are.

Niche overlap inferred from food webs is a necessary but not a sufficient condition for exploitation competition when one common limited resource is food. Niche overlap is neither necessary nor sufficient for interference competition.

CHAPTER THREE

Which Food Webs
Are Interval?

Frogs Eat Butterflies. Snakes Eat Frogs.
Hogs Eat Snakes. Men Eat Hogs.
Title of a poem by Wallace Stevens

3.1 SOURCES OF DATA

Thirty-one food webs from 22 different papers form the empirical basis of this study (Table 4). The reason for omitting source food webs, such as those of Root (1975) and Birkeland (1974), is given in section 2.2. We included source food web number 13, from which we draw no conclusions, out of sheer curiosity.

We omit hypothetical or schematic constructions: among others, Shelford's (1913, pp. 70, 167) for pond and prairie communities; Lindeman's (1942, p. 401) for a lake; Olson's (1966, p. 296) for Permian communities in northern Texas. Our justifications for omitting these webs are remarks such as "we have chosen a number of prairie animals and constructed them into an arbitrary community" (Shelford, 1913, p. 167) and "It [the reconstructed Permian food web] is, of course, overly simple as compared to conditions that must have actually existed" (Olson, 1966, p. 296).

We omit avowedly incomplete or tentative food webs. Though there may be no such thing as a complete food web, we exclude cases where the author reports that he has not described all the feeding relations he knows to be occurring. For example, Kashkarov and Kurbatov (1930, p. 37) comment, "The expedition worked only one month and was all the time en route without remaining long in one place. This influenced the results obtained." Carlson (1968, p. 167) says that the

28

TABLE 4. Code numbers, sources, and brief titles of the food webs used as data.

Code[a] number	Author[b]	Date	Page	Community	Type[c] of web
1.1	Bird	1930	383	Prairie, Canada	Community
1.2	Bird	1930	393	Willow forest, Canada	Community
1.3	Bird	1930	406	Aspen forest, Canada	Community
1.4	Bird	1930	410	Aspen parkland	Community
2	Clarke et al.	1967	1384	Sand bottom	Community
4	Hairston	1949	68	Salamanders, Appalachians	Sink
5	Hardy	1924	opp. 34	Herring, east coast of England	Sink
7	Koepcke	1952	24a	Sandy beaches, Peru	Community
8.1	Kohn	1959	70[d]	Conus, Hawaiian marine benches	Sink
8.2	Kohn	1959	74-75[d]	Conus, Hawaiian reef, benches and deep water	Sink
8.3	Kohn	1959	74-75[d]	Conus, Hawaiian subtidal reefs only	Sink
10	B. Menge & Mauzey	1968	MS[e]	Pisaster and Leptasterias	Community
11	Niering	1963	157	Kapingamarangi atoll, Carolines	Community
12	Paine	1963	67	Gastropods, Florida	Sink
13	Richards	1926	263	Pine feeders, Oxshott Heath	Source
15	Summerhayes & Elton	1923	232	Bear Island	Community
16.1	Paine	1966	67	Starfish, Mukkaw Bay, Wash.	Sink
16.2	Paine	1966	68	Starfish, Gulf of California	Sink
16.3	Paine	1966	69	Gastropods, Costa Rica	Sink
18	Minshall	1967	148	Morgan's Creek, Kentucky	Community
19	Valiela	1969	225	Dung	Sink
20	Stone	1969	459	Chaetognatha, Agulhas	Sink
23	Reynoldson & Young	1963	183	Lake-dwelling triclads	Sink
24	Teal	1962	616	Salt marsh, Georgia	Community
25	Harrison	1962	61	Rain forest, Malaysia	Community
26	Thomas	1962	185	River Teify, West Wales	Sink
27	Hartley	1948	7	Fish, River Cam, England	Sink
28.1	Fryer	1957	217	Rocky shore, Lake Nyasa	Community
28.2	Fryer	1957	218	Sandy shore, Lake Nyasa	Community
28.3	Fryer	1957	219	Crocodile Creek, Lake Nyasa	Community
29	Parsons & LeBrasseur	1970	341	Strait of Georgia, B.C.	Sink

[a] The number to the left of the decimal point identifies the source paper. Where there are several food webs in the same paper, or where more than one version of a single food web from a paper is analyzed, the first number to the right of the decimal point identifies the food web. Multiple versions of a food web, when they exist, are identified by the second number to the right of the decimal point.

[b] Full citations appear in the References.

[c] Community, sink, and source food webs are defined in section 2.2.

[d] Supplemented and corrected by personal communication.

[e] Personal communication.

"principal channels in nutrient and energy flow are tentatively diagrammed" in his food web, for which his caption reads "probable nutrient and energy flow." Clapham (1973, p. 113) describes his food web of a small meadow as a "diagrammatic representation, highly simplified." Pianka (1974, p. 219) reports "part of the food web in an Australian sandy desert . . . . A more detailed food web would separate all food types into species." Hiatt and Strasburg (1960, p. 126) comment, "only the major food pathways are shown in this foodweb, this being done for simplicity's sake because of the large number of species studied." Among these omitted food webs, some are interval and some are not. We consider the data sufficiently questionable that these results are uninformative.

We omit the food web of Petipa, Pavlova, and Mironov (1970) because the aggregation of kinds of organisms (e.g. herbivores, mixed food consumers, primary carnivores) was excessive for the purposes of the present analysis, although such aggregation was entirely appropriate for their study of mass and energy flows.

We probably include some food webs that ought to be discarded. Koepcke (1952, p. 2) characterized his food web as "eine schematische Darstellung" without specifying whether that included every trophic interaction observed. Minshall (1967, pp. 147, 148) says his food web shows the "principal" or "major" pathways of energy flow. Stone (1969, p. 459) tabulates the "principal prey . . . in an abridged fashion."

The reader is invited, therefore, to apply his own evaluation of the data to the conclusions reached from them. We believe that the studies based on intensive observation of relatively few kinds of organisms, particularly those based on the exhaustive observation of diet of one or a few predators, provide more credible data than reports of wider scope such as those of Koepcke (1952) and Summerhayes and Elton (1923).

We study more than one version of many of the food webs. For example, three versions of the food web of Hardy (1924) are analyzed. One version (code number 5.11) contains the information observed by A. C. Hardy and indicated by solid arrows in his figure (1924, opposite his p. 34), as well as predation definitely mentioned in the text. This version indicates that whitebait herring (42 to 130 mm long) eat young sand eels (*Ammodytes*) and *Apherusa* (Hardy, 1924, pp. 12–13), although Hardy omits this information from his food web graph. A second version (code number 5.12) includes everything in the first, plus information from other research that Hardy's food web graph indicates by a broken arrow. A third version (code number 5.13) indicates in addition likely predation inferred from the text. Ambiguities in the art work of Hardy's graph have been resolved by plausible guesses.

So that our interpretations of text and figures can be checked and improved, all versions of the food webs are presented in Appendix 1.

3.2 REVIEW OF FOOD WEBS

We now review the food webs listed in Table 4 case by case.

In the aspen parkland of Central Canada, Bird identified three "major biotic communities": the prairie community (food web 1.1), willow communities (food web 1.2), and aspen communities (food web 1.3).

The differences between Bird's figure (1930, p. 383) for the prairie community and our food web matrix 1.1 illustrate once again how we have modified some of the graphically reported food webs. Bird (1930, p. 379) says frogs sometimes eat small individuals of their own species. Though he omits this predation from his figure (1930, p. 383), the intersection of the row and column [14] corresponding to frogs in Appendix 1 has a 1. Bird (1930, pp. 378, 380) says that the crow [13] takes whatever it happens to find, including waste grain, and that the horned lark [12] eats weed seeds. On the possibility that these seeds may be among those enumerated as [8],

Agropyron, Stipa, and *Helianthus,* we insert -1 in row [8] in columns [12] and [13]. The food web obtained by setting each -1 to 0 is food web 1.11, while the food web obtained by setting each -1 to 1 is food web 1.12.

For all versions, each single-habitat food web is interval. When the three communities are composed into a single schema of "biotic interaction" of the communities, the niche overlaps no longer can be represented in a single dimension. This is true for both versions 1.41 and 1.42 of the composite food web.

In this case, the food webs of the single habitats and the composite community are reported by the same observer. The differences between food webs 1.1 to 1.3 and food web 1.4 cannot be attributed to different definitions of "kind of organism" or to different intensities of observation.

Another comparison of single habitats and composite communities is possible with the data of Kohn (1959). Food web 8.1 gives the prey organisms consumed by vermivorous species of the gastropod genus *Conus* at marine bench stations in Hawaii, based on Kohn's Table 8 (1959, p. 70) and on subsequent additional data that Kohn has kindly made available privately (personal communication, 1976). Tabulated are the numbers of individuals of the polychaete prey species found in alimentary tracts of the *Conus* species. This matrix becomes the usual food web matrix when each positive number is replaced by 1. The resulting niche overlaps are consistent with a one-dimensional niche space.

Food web 8.2, based on Kohn's Table 13 (1959), plus subsequent additional data, reports feeding records of the same kind as food web 8.1, from subtidal reef stations and from marine bench and deep water habitats. It is not interval. In the top line of this table, Kohn reports the numbers of specimens he examined of each species of predator. These range from 4 to 342 specimens. It seems plausible that, when only a few specimens of a predator are examined, some kinds of prey eaten on occasion might not be seen. The resulting omission of some niche overlaps may cause a true underlying one-dimensional niche space to appear to be more than one dimensional. In a second analysis of food web 8.2, only predators represented by more than 20 specimens were included; this threshold was determined before knowing the results of the following calculations. The resulting niche overlap graph is still not an interval graph.

Food web 8.3 is also based on Kohn's Table 13 and subsequent data but includes only the specimens taken at subtidal reef stations. Marine bench and deep water habitats are excluded. Once again the niche overlap graph is not an interval graph. However, if one also excludes predators represented by 20 or fewer specimens taken at the subtidal reef stations, the resulting niche overlap graph *is* an interval graph.

In this case, restricting attention to the adequately sampled predators is not enough to make a composite community containing subtidal reef, marine bench and deep water habitats compatible with a one-dimensional niche space, but does yield a possibly one-dimensional niche space in a single habitat.

The dominant clique matrix of food web 8.3, including all predators whether adequately sampled or not, would have the consecutive 1's property if the diet of *Conus imperialis* overlapped each of *C. sponsalis*, *C. miles*, and *C. rattus*. Because it is rare, *C. imperialis* was sampled only 31 times in Hawaii. Some indirect evidence about what might be found in a larger sample is available from 32 specimens that Kohn examined in the Seychelles Islands (personal communication, 1976). There he found definite evidence of dietary overlap of *C. imperialis* with *C. rattus*, and a probable dietary overlap with *C. sponsalis* on species of the worm genus *Eunice*. The single-habitat sink food web would be interval if both of these overlaps were found in Hawaii and if *C. miles*, sampled only 11 times in Hawaii, also overlapped *C. imperialis*. These conditional statements may be interpreted as predictions to be tested by future Hawaiian field studies.

Koepcke's (1952) food web of the sandy beaches of Peru distinguishes between feeding relations based on observation (food web 7.11) and probable feeding relations (food web 7.12). Both the solid (observed) and broken (probable) arrows in the graphic food web suffer from severe ambiguities. We have little confidence in having always guessed right in constructing our food web matrix. Food web 7.11 includes observations only. Food web 7.12 includes observed and probable relations. Neither version is interval. The habitats in this food web include sea, land, and air, plus kinds of organisms that are encountered primarily north of 6° latitude and kinds of organisms encountered primarily south of 6° latitude. It seems natural to interpret this food web as describing a composite of different habitats. The finding that it is non-interval is therefore consistent with expectations and with the previous examples.

The food web of Summerhayes and Elton (1923, p. 232) also distinguishes observed feeding relations (solid arrows) from relations that are "probable, but no evidence from here." As in Koepcke's food web, there are substantial ambiguities in the graphic presentation. Our best guesses about the observed feeding relations are presented in food web 15.11. The observed and probable relations are in version 15.12. The entries of −2 in the matrix represent the use of moss, a food for some species on Bear Island, as a nest material by glaucous gulls and kittiwakes (Summerhayes and Elton, 1923, p. 221). These entries of −2 are replaced by 0 in versions 15.11 and 15.12 and by 1 in version 15.13.

Although the food web nominally pertains to Bear Island, the text reveals that "data from Spitsbergen animals of the same species are included" (p. 231). Some kinds of organisms appear more than once in the food web graph, but in different

34

habitats. For example, protozoa appear as land organisms, as fresh water plankton, and as fresh water bottom and littoral organisms. Since the distinction between these habitats appears at the outset of the report and is indicated graphically in the food web, we distinguish these organisms as different kinds. They are likely to be different collections of species.

All three versions of the food web are non-interval. This outcome is to be expected because the food web describes a composite community. However, since the expedition that produced the data spent a total of only 10 days on Bear Island, the sampling of diets may be less than exhaustive, and actual niche overlaps might be omitted. No quantitative data are offered.

Fryer (1959) reports feeding relations in three littoral habitats of Lake Nyasa: the rocky shore (food web 28.1), the sandy shore (food web 28.2), and Crocodile Creek (food web 28.3). These habitats are initially distinguished by physiographic features (pp. 157–61). All three food web matrices incorporate extensive additions to the relations reported in Fryer's food web graphs, based on observations reported in the surrounding text. Though no arrows appear in his graph, the lines are evidently assumed to point upwards. In the food web on the rocky shore (p. 217), the line from Elmid larvae to *Pseudotropheus fuscoides* should rather be directed to *Haplochromis ornatus* (Fryer's Table 6 and p. 180). Fryer's indication that *P. fuscoides* eats terrestrial insects with a question mark is indicated by the -1 in row [28] and column [14] of food web matrix 28.1. Version 28.11 replaces this -1 by 0, version 28.12 by 1. Both versions are interval, as expected for a single habitat.

Neither of food webs 28.2 and 28.3 is interval. In the present sample, these represent the only food webs evidently describing single habitats for which all versions (here only one each) of the food web are not interval. Unlike Kohn, Fryer presents no quantitative data about the number of specimens of each predator examined, so it is impossible to exclude from Fryer's food webs any predators that may have been lightly sampled.

Niering (1963, p. 157) reports the energy flow in the Kap-ingamarangi Atoll, including both the marine and terrestrial habitats. The food web matrix incorporates feeding relations based on his text; skinks and geckos are separated. Version 11.12 allows the possibilities that skinks eat geckos and that the reef heron eats fish, while version 11.11 does not. Version 11.11 is interval, while version 11.12 is not. Since the food web describes a composite community, the expectation that it be interval is not very strong, but it is not impossible that version 11.12 is wrong.

All the remaining 18 food webs listed in Table 4 are interval. Richard's (1926, p. 263) "food-cycle on young pine" (food web 13) is a source food web, and hence not informative about the community. We now review the remaining 17 food webs.

Clarke, Flechsig, and Grigg (1967, p. 1384) report the food web of the sand bottom community and attracted fauna, including feeding relations based only on direct observation (their solid lines, our version 2.11), plus these and relations based on inference (their dotted lines, our version 2.12).

Hairston (1949, p. 68) reports the identifiable stomach contents of 10 specimens of each of four species of the plethodontid salamanders, genus *Desmognathus*. His Table 13 arranges the salamanders from left to right in order of increasing terrestriality and the prey organisms from top to bottom in order of increasing terrestriality. Predatory relations cluster around the principal diagonal of the table.

Hardy (1924) reports the relation of the herring to the plankton community as a whole off the east coast of England. His food web graph contains ambiguities caused by arrows that end in mid-air. The three versions of food web 5 have already been described.

Food web 10 consolidates B. Menge's (1970) report of the food web in which the starfish *Leptasterias hexactis* is the top consumer and Mauzey's (1967) report of the food web for which the starfish *Pisaster ochraceus* is the top consumer. By oversight, additional data of Robert T. Paine (Dayton, 1973, p. 664) on the diet of *Pisaster* are omitted from food web 10. According to Paine, *Mytilus californianus* and *Chthamalus dalli* each contribute more than 10% of the prey items in the diet of *Pisaster*. Food web 10 shows no predation by *Pisaster* on *C. dalli* and omits *M. californianus* as a prey item. Otherwise Paine's data are consistent with food web 10. Since three items of prey are taken by all six predators in food web 10, the conclusion that the food web is interval remains unchanged. The statistical properties of the food web (Chapter 4) would be changed slightly by incorporation of these data. That different habitats have been compounded in the construction of food web 10 shows that a web of a composite community can (though it need not) be interval.

Paine (1963, p. 67) tabulates the diets of eight sympatric predatory gastropods observed on a Florida sand bar. His food web graph (p. 70) is an incomplete extract from his tabulation and is ignored here. The abundances observed during eight extensive censuses varied substantially from one species to another, but at least 26 specimens of the least abundant species were observed.

Paine (1966) graphs a north temperate sink food web at Mukkaw Bay (food web 16.1), a subtropical sink food web in the northern Gulf of California (food web 16.2), and a tropical sink food web in Costa Rica (food web 16.3). In transforming the second food web graph into food web matrix 16.2, arrows from *Cantharus* and *A. angelica* to *Heliaster* were added in view of Paine's (1966, p. 69) remark that *"Heliaster* consumes all other members of this subweb."

Minshall (1967, p. 148) summarizes the food web of Morgan's Creek, Kentucky, without quantitative information to indicate what might have been omitted. Although his study reveals the importance of leaf litter that falls into the stream as a source of food for the animals in the stream, his is not a source food web because he did not set out to observe what feeds on this litter, and did not exclude organisms (diatoms) that are primary producers.

Valiela (1969, p. 225) tabulates the food habits of "prominent members of the dung community" on the basis of feeding experiments and laboratory and field observations. Where Valiela indicates that an animal is known to feed, our food web matrix 19 shows 1. Where he indicates that an animal is "presumed to feed" or that "feeding [is] possible" our food web matrix shows a -1. Where Valiela indicates that an animal is "presumed not to feed" or "known not to feed" or where he leaves a blank, our food web matrix shows a 0. Version 19.11 replaces each -1 by 0; version 19.12 replaces each -1 by 1. Both adults and larvae of the two species of *Sphaeridium* are omitted since they are eaten by no other species and eat only dung; *Atheta* and *Aleocharine* B65 are omitted as predators because they are reported as eating nothing at all. The remaining six predators have a complete niche overlap graph since, among other items in their diet, they are all reported as sharing a single food source, dung.

Stone (1969, p. 459) tabulates "in an abridged fashion" food eaten by six chaetognath species. The entries in his table are described as percentages. Since neither the rows nor the columns nor the whole matrix sums to 100% it is difficult to understand what the percentages refer to. Food web matrix 20 reproduces his table.

Reynoldson and Young (1963, p. 183) report the kinds and quantities of prey eaten by four species of triclads in laboratory experiments. All four species ate an arthropod, *Diura bicaudata*. So the niche overlap graph is a complete graph.

Teal (1962, p. 616) gives a food web graph of a Georgia salt marsh.

Harrison (1962, p. 61) diagrams food chains in the Malaysian tropical rain forest. Several feeding relations mentioned in his text (p. 62) appear in food web matrix 25 in addition to those Harrison pictures. Predation by unidentified "attendant predators" is omitted from our matrix. Since Harrison (p. 60) says the rain forest comprises six "communities," ranging from the upper air community down to small ground animals, it may be surprising that the niche overlap graph of this food web is an interval graph. At the same time, the kinds of organisms identified in the food web are highly aggregated so that details of diet are not available. A single unexplained broken arrow in the diagram is omitted in version 25.11 and taken as indicating predation in version 25.12.

Thomas (1962, pp. 184–85) reports the food of brown trout, salmon, and eel in West Wales. Because there are only three predators in this sink web, its niche overlap graph is necessarily an interval graph. (The smallest non-interval graph has four points.) Food web matrix 26 is based on his Figure 3, in which food items are grouped together in major classes. This grouping makes apparent niche overlap even more likely. However, his Figures 2 and 4 show many specific prey items that all three predators take in common. The complete niche overlap graph is genuine.

At the other extreme from the aggregated reporting of Harrison and Thomas, Hartley (1948, pp. 7–9) tabulates in detail the numbers of fish in which each kind of food was found and the total number of prey specimens found in those fish, for each of 11 predatory fish and 84 kinds of food in a stretch of the River Cam, England. Food web matrix 27 reproduces the numbers of prey specimens. The five plants at the bottom of his tabulations are omitted, since here "numbers of prey specimens" has no meaning. As the niche overlap graph is the complete graph on 11 points, even without the plants, including them would not alter the results of the test for an interval graph.

Parsons and LeBrasseur (1970, p. 341) report a "tentative" food web graph for Saanich Inlet, based on both feeding experiments and field observations. Some of the arrows in their graph are solid and some are broken. No explanation of the difference is offered. Food web matrix 29 represents the relations implied by all the arrows as 1.

3.3 SUMMARY

Thirty-one food webs are tested to see if the niche overlaps they imply can be represented in a one-dimensional niche space. These consist of one source food web, 16 sink food webs, and 14 community food webs. The source web is interval, but gives no information about whether the community web of which it forms part is interval.

Of the 16 sink webs, 14 may describe single habitats and two may describe composites of more than one habitat, though the distinction is not always clear. Thirteen of the 14 single habitat sink webs are interval, and one of the two composite habitat sink webs is interval. When lightly sampled predators are excluded from the single habitat sink food web that is not interval, it becomes interval. The same exclusion does not change the non-interval composite-habitat sink food web to interval.

Of the 14 community food webs, nine describe single habitats and five composites of more than one habitat. Seven of the nine single habitat community food webs are interval; quantitative data regarding the intensity of sampling are not available for the two exceptions. One or two (depending on the version used) of the composite habitat community food webs are interval.

CHAPTER FOUR

Statistics of Food Webs

The more flesh, the more worms.
Hillel, *Ethics of the Fathers*

4.1 WHY STATISTICS OF FOOD WEBS?

If every (or nearly every) possible graph on n points were an interval graph, no one would be surprised if every (or nearly every) observed niche overlap graph on n points were an interval graph. Should one be surprised by the finding in the preceding chapter that nearly every observed single-habitat food web is interval?

Before the question can be answered, it is necessary to estimate in some way the universe from which the observed food webs may be drawn. A useful first step is to characterize their variability or statistics.

An analogy helps. Whether or not a food web is interval is a dichotomous attribute, like whether the traffic in a city drives on the left or the right. Suppose one observes, as is indeed likely, that the traffic in London, Brighton, and Southampton (England) drives on the left. If one took the universe of cities from which these came to be those in southern England, one would conclude that such a practice is unsurprising. If one expanded one's universe of cities to include those in Wales and Scotland, one would remain persuaded that such behavior is unsurprising and even is found in cities where other languages than English (such as Gaelic) are spoken. If one's universe of cities extended to all those in Europe or in the world excluding China, one would, on the contrary, find the practice in London, Brighton, and Southampton extraordinary.

We consider seven possible universes, or statistical models, for the structure of food webs. We compare the observed char-

TABLE 5. Statistics of the food web versions.

Code[a] Number	Rows[b]	Columns[c]	Interval[d]	Row ave.[e]	Row var.[f]	Row x^{2g}
1.11	10	14	1	2.600	3.600	19.343
1.12	10	14	1	2.800	5.111	26.735
1.21	8	8	1	1.875	0.411	3.287
1.22	8	8	1	2.375	0.554	3.007
1.30	19	22	1	2.105	4.433	64.427
1.41	28	25	0	2.107	2.692	58.100
1.42	28	25	0	2.179	3.115	63.110
2.11	10	11	1	3.000	1.054	9.485
2.12	11	18	1	3.545	3.473	13.378
4.00	40	4	1	1.375	0.446	49.811
5.11	23	5	1	1.478	0.806	38.817
5.12	36	18	1	2.361	3.552	83.204
5.13	36	18	1	2.444	3.683	81.517
7.11	41	50	0	4.439	21.402	222.320
7.12	42	58	0	4.643	21.991	220.100
8.10	21	9	1	2.190	1.362	23.058
8.20	33	13	0	2.394	2.059	45.076
8.30	31	13	0	2.129	1.716	42.863
10.11	20	6	1	2.450	2.787	44.756
10.12	20	6	1	2.500	2.895	45.574
11.11	18	19	1	2.222	2.183	27.430
11.12	18	19	0	2.333	2.353	27.157
12.00	32	8	1	2.125	1.488	41.620
13.00	8	9	1	1.875	2.982	23.366
15.11	23	24	0	2.348	2.146	31.072
15.12	24	27	0	2.792	3.737	42.504
15.13	24	27	0	2.875	4.114	44.826
16.10	6	2	1	1.333	2.667	6.000
16.20	12	7	1	2.500	1.909	16.085
16.30	4	2	1	1.500	0.333	4.000
18.00	8	11	1	4.500	1.714	4.613
19.11	15	6	1	1.933	1.924	31.294
19.12	29	6	1	2.793	1.241	26.356
20.00	8	6	1	2.875	4.411	23.009
23.00	8	4	1	2.500	1.429	12.052
24.00	7	5	1	1.429	0.286	4.152
25.11	5	8	1	2.800	3.200	8.186
25.12	6	8	1	2.500	3.100	11.254
26.00	14	3	1	2.714	0.374	19.111
27.00	84	11	1	3.143	6.341	263.890
28.11	14	28	1	6.714	29.143	74.520
28.12	14	28	1	6.786	28.951	73.473
28.20	18	34	0	4.333	11.059	52.294
28.30	15	28	0	3.467	16.695	85.985
29.00	6	5	1	1.667	0.667	5.431

[a] Code number = food web number keyed to Table 4, and version.

[b] Rows = number of kinds of prey preyed on by at least one predator = rows in the food web matrix.

[c] Columns = number of kinds of predators that prey on at least one kind of prey = columns in the food web matrix.

[d] Interval graph = 1 if the niche overlap graph is an interval graph; 0 otherwise.

[e] Row average = average number of predators per prey = average row sum = average number of 1's per row.

[f] Row variance = variance of the number of predators per prey = variance of the row sums = variance in the number of 1's per row.

[g] Row chi-squared = statistic to test whether row sums have 0-truncated binomial distribution (section 4.2).

Table 5 (Continued)

Row $P^{[h]}$	Col. ave.$^{[i]}$	Col. var.$^{[j]}$	Col. $X^{2[k]}$	Col. $P^{[l]}$	Type$^{[m]}$	Overlap$^{[n]}$
0.02200	1.857	0.901	13.158	0.43600	1	34
0.00200	2.000	1.077	13.590	0.40300	1	44
0.85700	1.875	1.268	10.146	0.18000	1	7
0.88400	2.375	1.696	9.214	0.23800	1	10
0.00000	1.818	1.534	36.842	0.01800	1	55
0.00100	2.360	3.740	57.902	0.00100	1	59
0.00100	2.440	3.840	56.166	0.00100	1	66
0.39400	2.727	2.218	13.443	0.20000	1	23
0.20300	2.167	1.912	26.915	0.05900	1	46
0.11500	13.750	0.917	0.305	0.95900	2	6
0.01500	6.800	23.600	19.851	0.00100	2	9
0.00001	4.722	9.977	42.826	0.00100	2	70
0.00010	4.889	11.752	48.729	0.00010	2	72
0.00000	3.640	15.174	247.680	0.00000	1	310
0.00000	3.362	13.989	293.450	0.00000	1	373
0.25894	5.111	24.111	50.833	0.00000	2	21
0.06248	6.077	28.077	68.562	0.00000	2	36
0.06024	5.077	17.577	50.879	0.00000	2	34
0.00100	8.167	51.767	53.587	0.00000	2	15
0.00100	8.333	49.867	51.306	0.00000	2	15
0.05200	2.105	5.655	83.499	0.00000	1	33
0.05600	2.211	5.620	75.862	0.00000	1	38
0.09600	8.500	22.000	24.685	0.00100	2	22
0.00100	1.667	2.000	23.617	0.00300	3	17
0.09465	2.250	2.804	45.889	0.00309	1	50
0.00792	2.481	3.875	60.421	0.00015	1	79
0.00417	2.556	3.872	57.589	0.00035	1	88
0.30600	4.000	8.000	6.085	0.01400	2	1
0.13800	4.286	17.905	40.261	0.00000	2	15
0.26100	3.000	2.000	2.769	0.09600	2	1
0.70700	3.273	1.618	9.083	0.52400	1	36
0.00500	4.833	26.567	41.371	0.00000	2	15
0.55300	13.500	152.700	105.810	0.00000	2	15
0.00200	3.833	0.967	2.512	0.77500	2	15
0.09900	5.000	4.000	6.429	0.09300	2	6
0.65600	2.000	2.000	8.382	0.07900	1	3
0.08500	1.750	0.500	5.168	0.64000	1	18
0.04700	1.875	0.411	3.524	0.83300	1	18
0.12000	12.667	1.333	2.211	0.33100	2	3
0.00000	24.000	116.200	67.783	0.00000	2	55
0.00000	3.357	6.682	78.386	0.00000	1	256
0.00000	3.393	6.766	78.483	0.00000	1	256
0.00010	2.294	4.881	113.180	0.00000	1	168
0.00000	1.857	1.534	44.509	0.01800	1	165
0.36600	2.000	1.000	4.396	0.35500	2	4

[h] Row P = significance level of the statistic in column 7; 0.00000 means less than 0.000005.

[i] Column average = average number of prey per predator = average column sum = average number of 1's per column.

[j] Column variance = variance of the number of prey per predator = variance of the column sums = variance in the number of 1's per column.

[k] Column chi-squared = statistic to test whether column sums have 0-truncated binomial distribution (section 4.2).

[l] Column P = significance level of the statistic in column 11; 0.00000 means less than 0.000005.

[m] Type = 1 if a community food web; 2 if a sink food web; 3 if a source food web (section 2.2).

[n] Overlap = number of niche overlaps implied by the food web matrix = number of edges in the niche overlap graph.

acteristics (Table 5) of the food webs in our sample with the characteristics of the graphs in these universes.

4.2 THE FOOD WEB MATRIX

4.2.1 *Are Predators and Prey Fussy?*

In a first model of a food web, let us suppose that every predator decides whether it will eat each kind of prey with some fixed probability p, which is the same for all predators and prey in that food web. Each predator is assumed to make each decision regarding each kind of prey independently of its other decisions and of every other predator's decisions.

No field biologist would believe such a model for an instant. Doubtless neither do any of the organisms they study. Such disbelief is not an absolute argument against organisms behaving in aggregate as if such a model were true. Let the data speak.

Under this model, the number of prey eaten by a given predator and the total number of prey eaten are binomially distributed. Our matrices exclude predators that take no prey among the organisms in the food web. So the column sums or numbers of kinds of prey taken by each predator should be 0-truncated binomially distributed (Johnson and Kotz, 1969, pp. 73–76). Let m be the number of rows of a food web matrix, n be the number of columns. Let "column variance" be the sample variance of the n column sums of the food web matrix, and "column average" be the sample mean of the n column sums. Then the same logic used to derive the variance test for homogeneity of the binomial distribution (Snedecor and Cochran, 1967, pp. 240–41) implies that the statistic

$$X^2 = (n - 1)(\text{column variance})/(\text{estimated variance}) \quad (4.1)$$

should have the distribution of χ^2 with $n - 1$ degrees of freedom (df). Here

estimated variance
$$= mpq(1 - q^m)^{-1} - m^2p^2q^m(1 - q^m)^{-2} \qquad (4.2)$$

and p is the solution, obtained by numerical iteration, of the maximum likelihood estimation equation

$$\text{column average} = mp(1 - q^m)^{-1}, \qquad (4.3)$$

and $q = 1 - p$. Column average, column variance, and X^2 appear in Table 5 for each food web, along with the probability P of a value of X^2 larger than that observed occurring by chance alone, assuming the model were true.

If this model were always true, then in a sample of independently chosen food webs, the values of P would be uniformly distributed between 0 and 1. A frequency histogram of P would approximate a horizontal straight line. Fisher (1970, section 21.1) gives a formal test for whether a set of independent P values conforms to a uniform distribution. Because our P values arise in part from multiple versions of single food webs, we shall be content with an informal examination of the P values.

The actual frequency histogram of P (Fig. 7a) bears no resemblance to a horizontal straight line. In this and all subsequent frequency histograms, each version of a food web appears once. All of the non-interval food webs have P less than 0.05 (Fig. 7b). The interval sink food webs (Fig. 7c) also have a substantial proportion with P less than 0.05. The frequency histogram (Fig. 7d) of the interval community food webs approximates a uniform distribution better, though still not well. A 0-truncated binomial model for the column sums (number of kinds of prey per predator) of interval community food webs may not be a hopeless model.

With retrospective wisdom, one can argue that, if this first model describes any food webs at all, it is more plausible that it would describe community food webs than sink food webs. In community food webs, unlike sink food webs, there is no selection of one or more kinds of organisms precisely because

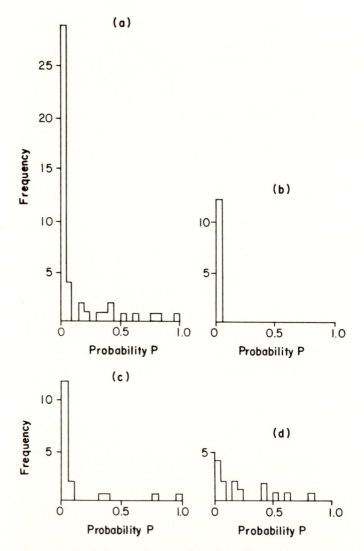

FIGURE 7. The frequency histogram of the probability P of the variance test, equation (4.1), of the fit of the number of kinds of prey per predator to the 0-truncated binomial distribution. (a) All food web versions. (b) Non-interval food web versions. (c) Sink food web versions which are interval. (d) Community food web versions that are interval. The histograms are far from a uniform distribution in (a), (b), and (c), but not in (d). (Based on Table 5.)

their diet is of interest, and hence no imposed heterogeneity in number of kinds of prey.

There is an interesting, though unflattering, formal analogy between this finding of homogeneity in the distribution of the number of kinds of prey per predator and the finding in hierarchical organizations of humans that each person at any given level is often responsible for the work of x people in the level below him. Here x usually lies between 3 and 10 at executive levels and does not vary much within a firm (Bartholomew, 1967, p. 193).

Now consider a second model of a food web. Suppose that every kind of organism is exposed to predation by each kind of organism with some fixed probability p, which is the same for all predators and prey in that food web (but not necessarily equal to p in the first model). Again each predatory relation is assumed independent of every other.

Under this model the number of predators on a given prey and the total number of predators are binomially distributed. Because our matrices exclude prey not eaten by organisms in the food webs, the row sums, or numbers of kinds of predators taking each prey, should be 0-truncated binomially distributed. The statistic obtained from equation (4.1) after interchanging n and m, and row and column, should have the distribution of χ^2 with $m - 1$ df.

The frequency histogram of P values for this test (Figure 8a) shows that the model is inadequate for the ensemble of food webs. The non-interval food webs all have low values of P (Figure 8b). Among the others, the community food webs (Figure 8d) have a histogram that approximates a uniform distribution more closely than that of the sink food webs (Figure 8c). Again, this difference is plausible. In a sink food web, the possible prey organisms include the kinds of organisms that are chosen as the sinks and the kinds of organisms that are the diet of those sinks. Hence the number of predators on a sink organism may or may not exceed 0, but the number of predators on a

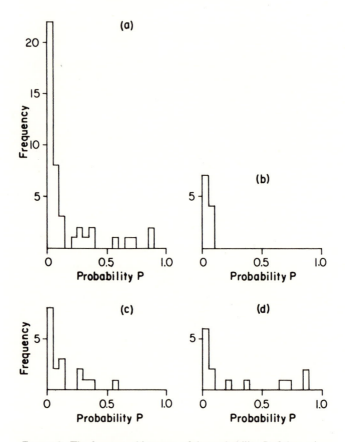

FIGURE 8. The frequency histogram of the probability P of the variance test, equation (4.1), of the fit of the number of kinds of predators taking each prey to the 0-truncated binomial distribution. (a) All food web versions. (b) Non-interval food web versions. (c) Sink food web versions which are interval. (d) Community food web versions that are interval. The histogram in (d) approximates a uniform distribution more closely than the histograms in (a), (b), and (c). (Based on Table 5.)

diet organism must be positive. In a community food web, there is no such imposed heterogeneity.

Of the 20 sink food web versions in Table 5, only two are non-interval. These two, food webs 8.2 and 8.3, are discussed

48

in detail in section 3.2. There are not enough non-interval sink food webs for meaningful comparisons between interval and non-interval sink food webs.

Among community food webs, those that are interval resemble those that are non-interval in several respects. The distribution of the average number of kinds of predators per prey, or row averages, among interval community food webs is not noticeably different from the distribution among non-interval community food webs (Table 6). The distribution of the average number of kinds of prey per predator, or column averages, among interval community food webs hardly differs from the distribution among non-interval community food webs.

TABLE 6. The distribution among community food web versions of the average number of predators per prey (row averages), the variance of predators per prey (row variances), the average number of prey per predator (column averages), and the variance of prey per predator (column variances), according to whether the version is interval or non-interval; based on Table 5.

	Interval (14 food web versions)	Non-interval (10 food web versions)
	Average predators per prey (row averages)	
Mean	3.2	3.2
S.d.[a]	1.7	1.0
Range	1.4 to 6.8	2.1 to 4.6
	Variance of predators per prey (row variances)	
Mean	6.2	8.9
S.d.	9.8	8.2
Range	0.3 to 29.1	2.1 to 22.0
	Average prey per predator (column averages)	
Mean	2.3	2.5
S.d.	0.6	0.5
Range	1.8 to 3.4	1.9 to 3.6
	Variance of prey per predator (column variances)	
Mean	2.4	5.9
S.d.	2.2	4.7
Range	0.4 to 6.8	1.5 to 15.2

[a] Standard deviation.

In other respects, community food webs that are interval differ from those that are non-interval.

Interval community food webs have fewer kinds of predators (Figure 9a,b) and fewer kinds of prey (Figure 9c,d) than non-interval community food webs. All but two of the single habitat food webs are interval (Chapter 3), and the multiple-habitat food webs have more kinds of predators and prey than do the single-habitat food webs. So in the present set of food webs, an

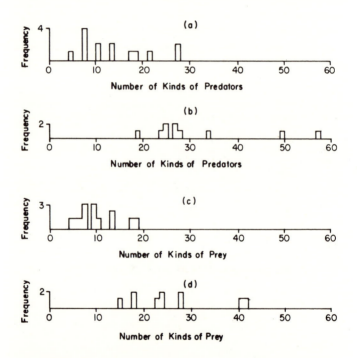

FIGURE 9. (a) The frequency histogram of number of kinds of predators in interval community food webs. (b) The frequency histogram of number of kinds of predators in non-interval community food webs. The number of kinds of predators is generally higher in (b) than in (a). (c) The frequency histogram of number of kinds of prey in interval community food webs. (d) The frequency histogram of number of kinds of prey in non-interval community food webs. The number of kinds of prey is generally higher in (d) than in (c). (Based on Table 5.)

interval food web is much more likely to describe a single habitat than a composite community. This helps explain the smaller numbers of kinds of predators and prey in interval community food webs.

Among non-interval community food webs, the average row variance is nearly 50% greater than the average row variance among interval community food webs. Among non-interval community food webs, the average column variance is more than twice the average column variance among interval community food webs. This shows that the higher X^2 statistics or lower P values for the 0-truncated binomial variance test among non-interval food webs is due to an increase in the variation within each food web in the column or row sums, rather than due to a change in the averages.

Predators in interval community food webs, when compared with those in non-interval community food webs, maintain roughly the same average number of kinds of prey taken in the face of a smaller number of kinds of prey available by taking a greater percentage of the available kinds of prey. This also reduces the variation among predators within a community in the number of kinds of prey taken. The same conclusion could be phrased in terms of prey.

This conclusion is empirical, not logical. As Figure 17 will show, the probability that a random graph is an interval graph at first declines as the ratio of number of edges (niche overlaps) to vertices (kinds of predators) increases, and then increases again. There is no logical reason why real interval niche overlap graphs should be associated with increasing use of prey, and presumably increasing niche overlap.

4.2.2 *The Ratio of Prey to Predators*

Figure 10 plots the number m of kinds of prey, or rows of the food web matrix, on the ordinate against the number n of kinds of predators, or columns of the food web matrix, on the abscissa, for each food web version in Table 5 except the single source

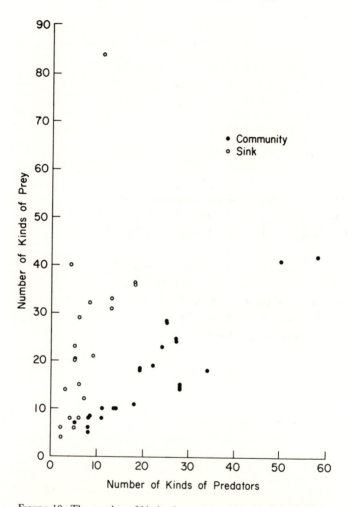

FIGURE 10. The number of kinds of prey (rows of the food web matrix) as a function of the number of kinds of predators (columns of the food web matrix). Community food web versions are represented by solid circles. Sink food web versions are represented by open circles. The source food web 13 is omitted. (Based on Table 5.)

food web, number 13. The community food webs are represented by darkened circles; the sink food webs, by open circles.
Several aspects of this graph are remarkable.

First, the distinction between community and sink food webs
cleanly separates the data points into two distinct groups. The
distinction arose and was applied to individual food webs in
advance of the preparation of Figure 10, for the different purposes described in Chapter 2. Figure 10 shows that the difference
in the way the two kinds of food web are composed makes a
consistent quantitative difference, not anticipated, in the food
web's structure. For any given number of predators, there are
more prey in a sink food web than in a community food web.
Retrospectively, it is easy to see why. A sink food web includes
all the prey of the chosen sinks, without necessarily including
all the other predators of those prey. Differently put, for a
fixed number of prey, a community food web has more predators than a sink food web because a community food web does
not selectively omit sinks (predators) other than a set of special
interest.

Second, in community food webs, the number of prey is very
nearly proportional to the number of predators. A least squares
regression of m against n gives

$$m = 1.79 + 0.71n. \tag{4.4}$$

The sample standard deviation of the regression coefficient is
0.07 and the linear correlation coefficient between m and n is
0.90. The standard error of estimate, or sample standard deviation from regression, is 4.62. As is obvious from Figure 10, the
regression may be well approximated by a straight line through
the origin. The least squares regression is then

$$m = 0.77n. \tag{4.5}$$

In order to avoid making any claims about the sampling of
food webs, we forego attaching probability values to any of

these statistics. Fortunately probability values are entirely un-necessary here because the picture is so clear.

The simple proportionality between the number of kinds of prey and the number of kinds of predators in Figure 10 and equation (4.5) is based on 24 versions of 14 different food webs reported by 9 different authors over a time span of decades. The food webs were collected and encoded with no anticipation that such a simplicity would emerge. It therefore seems likely that this invariance in the proportions of predators and prey represents a fact about nature, rather than an artifact of collusion or convention. The units of measurement here are the "kinds" of organisms ecologists record in food webs, rather than, say, species.

Third, knowing now that the proportion of prey to predators is a scale-invariant feature of community food webs, we can predict the proportion quantitatively from facts already at our disposal. Suppose a food web graph has A directed edges from prey to predators, where A is an integer between $\max(m, n)$ and mn. Equivalently suppose the food web matrix has A entries of 1 and $mn - A$ entries of 0. Then tautologously

$$A = (\text{column average}) \times n = (\text{row average}) \times m. \quad (4.6)$$

Now the mean over all 24 community food web versions, whether or not they are interval, of the column averages is 2.418 and of the row averages is 3.199. If we suppose that these means apply to an arbitrary matrix and substitute them into equation (4.6), we predict

$$m/n = 2.418/3.199 = 0.756, \quad (4.7)$$

which differs trivially from the least squares regression in equation (4.5).

The transparent simplicity of the argument from the pro-portionality between m and n to the prediction in equation (4.7) may raise a suspicion that its success depends on a fact of arithmetic rather than on the observed invariance of proportions of predators and prey in nature. A simple numerical example proves this suspicion false. Suppose our sample of community food webs consisted of just two food webs. Suppose the first food web matrix had $m_1 = 8$ rows, $n_1 = 6$ columns, and $A_1 = 19.2$ 1's (neglecting the requirement that A_1 be integer for the sake of argument). Then its row average$_1$ is 2.4 and column average$_1$ is 3.2. Suppose the second food web matrix had $m_2 = 4$, $n_2 = 10$, and $A_2 = 16$. Then row average$_2 = 4.0$ and column average$_2 = 1.6$. Then the mean over both food webs of the row averages is 3.2 and the mean of the column averages is 2.4, which are close enough to the observed. But the straight line through the pairs (n, m) satisfies $m = 14 - n$. Only because nature assures a constant proportion of prey kinds to predator kinds can we apply the mean row average and the mean column average to an arbitrary food web, and hence all food webs (Cohen, 1977b).

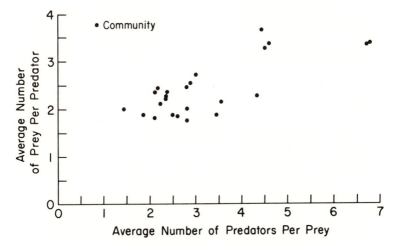

FIGURE 11. The average number of kinds of prey per kind of predator (the column average of a food web matrix) as a function of the average number of kinds of predators per kind of prey (the row average of a food web matrix) in community food web versions. (Based on Table 5.)

If the ratio of prey to predators in community food webs is a constant on the order of 3/4, we can make another prediction. Table 6 shows that both column averages and row averages do vary substantially from one food web to another. Dividing equation (4.6) by n leads to the prediction, if $m/n = 3/4$, that a regression of column average against row average should be a straight line through the origin with slope 3/4. Figure 11 shows the column average and row average of each community food web version. The regression coefficient of a straight line through the origin is 0.69, not far from 3/4.

The finding that in community food webs the ratio of prey to predators approximates 3/4 may prove useful to theoretical papers on whether there can be more predators than prey (Haigh and Maynard Smith, 1972).

4.2.3 *The Length of Food Chains*

A food chain of length k is an ordered set of $k + 1$ vertices (with repetitions possible) from a food web graph such that there is a directed edge in the food web graph from each vertex in the ordered set to the next, except for the last vertex. The length of a food chain is one less than the number of kinds of organisms in the chain (counting repetitions as distinct).

Cannibalism is the appearance of the same kind of organism as two or more consecutive vertices in a food chain. A loop is the appearance of the same kind of organism as two or more nonconsecutive vertices in a food chain. A loop would occur if, for example, A ate B, B ate C, and C ate A; or if A ate B and B ate A.

A maximal food chain in a food web is a food chain in which no kind of organism appears more than once and to which it is not possible to add any more kinds of organisms at either the beginning or the end and still have a food chain. For example, if (A, B, C, D) is a food chain of a food web, then neither (A, B, C) nor (B, C, D) is a maximal food chain, though (A, E, D) may well be. (A, E, C) would not be a maximal chain of this same food web because there is an arrow from C to D and (A, E, C) could be extended to (A, E, C, D). Any of these kinds of organisms A, B, C, D, E might be cannibals. The length of a maximal food chain cannot exceed the number of predators in its food web.

In the two interval food web versions and the two non-interval food web versions with the largest number of predators, there were no food chains with loops (Table 7) though there was one cannibal. The frequency distribution of the length of the maximal food chains shows that, in these food webs, the observed upper limit of food chain length was far less than the number of predators. Though two examples of each kind provide little basis for comparison, there are no striking differences in the frequency distributions between interval and non-interval food webs.

The absence of loops supports Gallopin's (1972, p. 266) generalization that "directed food webs are in general acyclic [without loops], although exceptions are possible."

Gallopin and many other sources attribute to Hutchinson (1959) the generalization that "in directed food webs, in general, there are not arc progressions (food chains) more than five links long." The 687 maximal food chains of length greater than 5, out of 2,164 maximal food chains, in Koepcke's (1952) food chain shows that this generalization is false in at least some food webs.

In fact Hutchinson offers no generalization like the one attributed to him. He defines an Eltonian predator chain (p. 147) as a food chain in which the first kind of organism is a green plant, the next a herbivorous animal, the next a primary carnivore, the next a secondary carnivore, and so on, and in which each predator is larger and rarer than its prey. He says, "Five animal links [or vertices, correspondingly to four directed edges] are certainly possible, a few fairly clear cut cases having been in fact recorded," but argues that 50 animal vertices in an Eltonian predator chain are impossible. He then generalizes without setting any specific quantitative upper limits: "Clearly the Eltonian food-chain of itself cannot give any great diversity, and the same is almost certainly true of the other types of food chain, based on detritus feeding or on parasitism." Table 7 provides no evidence against this generalization. A quantitative model to predict the frequency distributions in Table 7 would be desirable.

4.3 THE NICHE OVERLAP MATRIX

4.3.1 *Six Models Based on the Food Web*

Whether a food web is interval depends on whether its niche overlap graph is an interval graph. Since many different food webs are compatible with the same niche overlap graph, it is important to study niche overlap directly.

The two most elementary features of a niche overlap graph are the number n of vertices, equal to the number of predators in the food web, and the number E of edges. E depends on the number and arrangement of the A arrows or directed edges in the food web graph, or the A 1's in the food web matrix.

We now describe six models, which include the two presented in section 4.2.1, for the number and arrangement of 1's in the food web matrix. Comparison of the observed number E of edges in the niche overlap graph with the number expected from each of these six models shows that one and only one of these six models provides an excellent quantitative description of the number of overlaps among predators' niches.

Model 1: Fixed row sums. Take the number of 1's in each row of the food web matrix as given. Assume that these 1's are randomly distributed among the columns. Biologically, this model assumes that the number of predators on each prey must exactly equal the number observed, but that the identity of the predators on each prey is determined entirely at random.

Model 2: Truncated binomial row sums. Take the number of 1's in each row of the food web matrix as the mean of a

TABLE 7. Frequency distributions of lengths of maximal food chains in four food webs.

	Interval food webs		Non-interval food webs	
	1.3 Aspen forest, Canada	28.12 Rocky shore, Lake Nyasa	28.2 Sandy shore, Lake Nyasa	7.12 Sandy beaches, Peru
Predators	22	28	34	58
Length of chain[a]		Frequency		
1	2	20	21	1
2	9	55	37	62
3	6	34	11	234
4	7	0	0	460
5	0	0	0	720
6	0	0	0	561
7	0	0	0	126
>7	0	0	0	0
Number of chains	24	109	69	2164
Average length	2.75	2.13	1.86	4.86
Number of loops[b]	0	0	0	0

[a] Number of directed edges in the chain, or one less than the number of kinds of organisms in the chain.
[b] A loop is the appearance of the same kind of organism at two or more non-consecutive positions in a food chain.

0-truncated binomial distribution specific for that row. Conditional on the appearance of at least one 1 in the row, assume that each entry in the row is independently set equal to 1 with probability equal to the p parameter of the 0-truncated binomial distribution. Biologically, this model assumes that each prey has a characteristic risk of being preyed on, and that it is exposed to this risk independently for each predator. (This is the second model of section 4.2.1.)

Model 3: Fixed column sums. Take the number of 1's in each column of the food web matrix as given. Assume that these 1's are randomly distributed among the rows within a column. Biologically, this model assumes that the number of prey each predator takes must exactly equal the number observed, but that the identity of the prey is determined entirely at random.

Model 4: Truncated binomial column sums. Take the number of 1's in each column of the food web matrix as the mean of a 0-truncated binomial distribution specific for that column. Conditional on the appearance of at least one 1 in the column, assume that each entry in the column is independently set equal to 1 with probability equal to the p parameter of the 0-truncated binomial distribution. Biologically, this model assumes that each predator has a characteristic propensity to prey that it exercises independently on each kind of prey organism. (This is the first model of section 4.2.1.)

Model 5: Fixed matrix sum. Take the total number A of 1's in the food web matrix as given. Assume that these 1's are randomly distributed among the nm elements of the matrix. Biologically, this model assumes a fixed number of predator-prey pairs, with the identity of the predator and prey in each pair determined at random.

Model 6: Binomial matrix sum. Take A as the expectation of a binomial distribution of 1's in each food web matrix. Assume that each element of the matrix is independently set equal to 1 with probability $A/(mn)$, equal to 0 otherwise. In principle it is possible that this model would generate a food web matrix with no 1's, but, given the characteristics of the observed food webs, the very minor increase in precision that would be gained by insisting on 0-truncation has been sacrificed in favor of the analytical convenience of being able to assume that each element in the food web matrix is independent. Biologically, this model assumes that every predator in a given food web has a constant and independent probability of preying on every prey in that food web.

For each of the 45 food web versions in Table 5, the mean and variance according to each model of the number of edges in the niche overlap graph has been calculated (Table 8) according to expressions in Appendix 2. Models 1, 3, and 5, with fixed sums, are given in Table 8a; models 2, 4, and 6, with binomially distributed sums, are given in Table 8b.

Table 8 shows that the theoretical mean number of overlaps for model 1 is less than or equal to the mean for model 2, and the mean for model 4 is less than or equal to the mean for model 3, in every food web version. Proofs that these inequalities hold in general are given in Appendix 2. Of the 45 cases in Table 8, the mean for model 2 is greater than or equal to the mean for model 3 in 40 cases, for model 4 in 43 cases, for model 5 in 42 cases, and for model 6 in 43 cases.

The variance of model 3 is less than or equal to the variance of model 4, and the variance of model 5 is less than or equal to the variance of model 6 in all 45 cases. The variance of model 5 is strictly less than the variance of model 6 in 43 cases. Pathological counterexamples can be constructed to show that these inequalities are not always necessary, although we conjecture that the inequalities always hold for food web matrices with at least one 1 in each row and at least one 1 in each column. The variance of model 1 is less than or equal to the variance of model 2 in 44 cases, of model 5 in 41 cases, and of model 6 in 44 cases.

Assuming that the number E of edges is approximately normally distributed with the mean and variance given by each

TABLE 8. The observed frequency of niche overlap (number of edges) in the niche overlap matrix, the mean and variance of the frequency of niche overlap (number of edges) in the niche overlap matrix predicted by six models, and the probabilities under a normal approximation of deviations from the predicted means as large as or larger than those observed, for each food web version.

8a

Code[a] Number	Overlap[b]	Fixed row sums[c] Mean[f]	Variance[g]	p[h]	Fixed column sums[d] Mean	Variance	p	Fixed matrix sums[e] Mean	Variance	p
1.11	34	32.2740	5.3360	0.4549	28.3190	16.5450	0.1625	26.6600	15.4540	0.0619
1.12	44	42.3190	8.8320	0.5717	31.8170	16.8640	0.0030	30.3020	16.9390	0.0009
1.21	7	7.1570	0.6130	0.8414	10.5000	5.2050	0.1250	10.0840	5.1400	0.1737
1.22	10	12.0450	1.7180	0.1187	15.0890	5.2330	0.0261	14.7520	6.1410	0.0551
1.30	55	56.6450	8.1770	0.5652	37.0940	27.0230	0.0006	36.5190	26.2740	0.0003
1.41	59	63.0410	6.4130	0.1106	51.9640	35.2820	0.2362	53.8110	38.8910	0.4053
1.42	66	70.2950	8.8020	0.1477	55.0730	36.5820	0.0708	57.1950	40.7980	0.1680
2.11	23	29.6050	5.5390	0.0050	29.4960	9.7490	0.0375	29.8330	12.9870	0.0580
2.12	46	56.1910	10.6910	0.0018	53.8600	26.4520	0.1265	53.8050	30.7690	0.1594
4.00	6	5.8950	0.0992	0.7397	5.9960	0.0044	0.9468	5.9820	0.0183	0.8923
5.11	9	9.0550	0.8386	0.9519	7.2420	1.3660	0.1325	8.9700	0.9264	0.9755
5.12	70	86.5720	21.0920	0.0003	64.7800	27.9340	0.3234	71.2070	35.8640	0.8403
5.13	72	90.0680	22.0620	0.0001	66.6080	27.4160	0.3032	74.8810	36.4150	0.6331
7.11	310	566.0900	234.7500	0.0000	262.0900	133.5500	0.0000	338.8800	232.3700	0.0581
7.12	373	647.7500	245.4500	0.0000	308.9100	168.6600	0.0000	390.3600	280.1400	0.3000
8.10	21	25.5330	4.3190	0.0292	19.4850	4.9780	0.4793	26.3570	7.1470	0.0451
8.20	36	54.3070	10.5180	0.0000	39.5200	11.6790	0.3030	53.3960	17.1240	0.0000
8.30	34	44.6780	8.6580	0.0003	35.2460	12.4790	0.7243	44.6610	18.2760	0.0126
10.11	15	15.0000	0.0000	1.0000	12.2000	1.4258	0.0190	14.7321	0.2865	0.6167
10.12	15	15.0000	0.0000	1.0000	12.5376	1.3767	0.0359	14.7795	0.2361	0.6499
11.11	33	38.7250	4.2120	0.0053	32.7760	18.9790	0.9590	37.1470	24.8000	0.4050
11.12	38	42.6870	5.2050	0.0400	35.8590	20.1590	0.4768	40.5500	26.4890	0.6202
12.00	22	25.8490	1.9110	0.0054	22.8430	2.8900	0.6200	25.5760	2.3810	0.0205
13.00	17	16.1500	0.4750	0.2177	10.6070	5.5960	0.0069	10.5570	5.8320	0.0076
15.11	50	54.8360	5.0380	0.0312	52.8160	34.2890	0.6306	54.2800	37.9600	0.4872
15.12	79	91.3079	12.7500	0.0006	74.6680	44.6550	0.5169	79.3230	54.1220	0.9650
15.13	88	98.0820	15.4330	0.0103	78.8620	46.3560	0.1795	83.5760	56.3280	0.5556
16.10	1	1.0000	0.0000	1.0000	1.0000	0.0000	1.0000	1.0000	0.0000	1.0000
16.20	15	17.7940	2.3450	0.0681	12.7250	2.3350	0.1366	17.2800	3.3270	0.2112
16.30	1	1.0000	0.0000	1.0000	1.0000	0.0000	1.0000	1.0000	0.0000	1.0000
18.00	36	41.7880	8.9030	0.0524	44.2000	6.8960	0.0018	42.9340	12.6820	0.0515

Code[a]	Edges[b]	Model 1[c]			Model 3[d]			Model 5[e]		
		Expected[f]	Variance[g]	P[h]	Expected[f]	Variance[g]	P[h]	Expected[f]	Variance[g]	P[h]
19.11	15	15.0000	0.0000	1.0000	8.3810	1.8050	0.0000	12.3870	2.2010	0.0782
19.12	15	15.0000	0.0000	1.0000	11.4830	1.0700	0.0007	14.9950	0.0050	0.9442
20.00	15	15.0000	0.0000	1.0000	13.9500	0.8536	0.2558	13.4860	1.6290	0.2355
23.00	6	6.0000	0.0000	1.0000	5.8929	0.1033	0.7389	5.9660	0.0361	0.8574
24.00	3	2.7100	0.2259	0.5418	4.4000	1.8640	0.3052	4.4960	1.9860	0.2883
25.11	18	15.3720	2.7900	0.1157	15.0000	5.7000	0.2089	13.4800	5.7580	0.0596
25.12	18	15.3720	2.7900	0.1157	14.5000	6.0610	0.1551	12.9130	5.7160	0.0334
26.00	3	3.0000	0.0000	1.0000	3.0000	0.0000	1.0000	3.0000	0.0000	1.0000
27.00	55	55.0000	0.0000	1.0000	51.4710	2.1620	0.0164	54.9680	0.0325	0.8593
28.11	256	287.9700	143.8900	0.0077	178.6500	56.4130	0.0000	213.7000	119.6400	0.0001
28.12	256	289.1900	142.7600	0.0055	180.4100	56.1700	0.0000	216.7200	120.7100	0.0003
28.20	168	189.9200	51.0860	0.0022	130.4700	72.7910	0.0000	142.6100	95.5940	0.0094
28.30	165	163.4800	24.2040	0.7575	78.8600	52.2550	0.0000	77.4980	52.5880	0.0000
29.00	4	4.3300	0.4011	0.6023	5.3830	1.9510	0.3220	5.1270	2.0240	0.4282

[a] Code number of food web version, as in Table 5.

[b] Number of edges in the niche overlap graph.

[c] Model 1 in section 4.3.1.

[d] Model 3 in section 4.3.1.

[e] Model 5 in section 4.3.1.

[f] Expected number of edges in the niche overlap graph, according to this model, calculated according to Appendix 2.

[g] Variance of the number of edges in the niche overlap graph, according to this model, calculated according to Appendix 2.

[h] Assuming that the number of edges in the observed niche overlap graph is a normal random variable with the calculated mean and variance, P is the probability of a deviation from the mean at least as large in magnitude as the observed deviation.

TABLE 8. The observed frequency of niche overlap (number of edges) in the niche overlap matrix, the mean and variance of the frequency of niche overlap (number of edges) in the niche overlap matrix predicted by six models, and the probabilities under a normal approximation of deviations from the predicted means as large as or larger than those observed, for each food web version.

8b

Code Number	Overlap	Binomial row sums[i]			Binomial column sums[j]			Binomial matrix sums[k]		
		Mean	Variance	P	Mean	Variance	P	Mean	Variance	P
1.11	34	36.8450	142.2630	0.8114	26.9470	40.6480	0.2686	26.9370	81.0710	0.4328
1.12	44	46.3260	166.7430	0.8570	30.0470	43.7090	0.0348	30.5000	89.4490	0.1535
1.21	7	9.6260	19.1600	0.5486	10.0450	9.2430	0.3165	10.1820	18.9670	0.4650
1.22	10	14.8180	22.4310	0.3090	14.0160	11.3110	0.2325	14.6150	21.6760	0.3215
1.30	55	63.9720	428.1200	0.6646	35.9860	62.1930	0.0159	37.0450	131.7300	0.1177
1.41	59	75.4630	444.9400	0.4351	49.8800	91.1790	0.3395	54.2880	189.7200	0.7323
1.42	66	82.9000	486.0000	0.4433	52.7450	97.3310	0.1791	57.6600	202.5200	0.5578
2.11	23	33.5640	60.4700	0.1743	27.3620	25.5150	0.3878	29.6080	52.9510	0.3638
2.12	46	64.3910	255.7100	0.2501	51.0520	77.4040	0.5658	53.9950	189.8000	0.5617
4.00	6	5.9630	0.0446	0.8622	5.9580	0.0450	0.8427	5.9610	0.0419	0.8480
5.11	9	9.3970	1.2000	0.7167	6.8340	2.1270	0.1374	8.7800	1.6460	0.8639
5.12	70	96.0770	222.6200	0.0805	60.4780	85.0120	0.3017	71.0900	144.3500	0.9277
5.13	72	99.5770	210.4500	0.0573	62.2120	83.5280	0.2842	74.7180	146.0600	0.8221
7.11	310	600.0200	4342.6000	0.0000	252.5300	465.5900	0.0077	339.4100	1899.1000	0.4998
7.12	373	690.7600	6283.5000	0.0001	298.0900	626.9000	0.0028	391.1300	2485.9000	0.7161
8.10	21	28.4480	18.5450	0.0837	18.5290	9.2860	0.4173	26.0140	18.5930	0.2449
8.20	36	60.1790	52.9780	0.0009	37.2980	26.4720	0.8008	53.0160	51.6280	0.0179
8.30	34	51.2750	75.0200	0.0461	33.1970	27.8210	0.8790	44.4210	59.6700	0.1773
10.11	15	15.0000	0.0000	1.0000	11.7086	2.4552	0.0357	14.6094	0.5504	0.5986
10.12	15	15.0000	0.0000	1.0000	12.0322	2.4154	0.0562	14.6690	0.4590	0.6252
11.11	33	47.2260	225.1040	0.3275	32.2070	34.5740	0.8927	37.5490	123.0200	0.6817
11.12	38	51.5880	225.1500	0.3651	35.0070	39.4430	0.4766	40.9230	134.0000	0.8007
12.00	22	26.7990	2.4240	0.0021	21.6960	6.5110	0.9052	25.3070	4.4810	0.1183
13.00	17	17.6370	53.1450	0.9304	10.5230	8.4020	0.0255	10.7570	23.4820	0.1977
15.11	50	67.6670	348.9000	0.3442	50.6580	89.0770	0.9444	54.7640	199.8900	0.7362
15.12	79	107.2800	621.0300	0.2564	71.4130	127.5000	0.5016	79.8070	319.2500	0.9640
15.13	88	114.1800	649.3500	0.3043	75.2190	137.6100	0.2759	84.0360	336.2700	0.8289
16.10	1	1.0000	0.0000	1.0000	1.0000	0.0000	1.0000	0.9706	0.0285	0.8618
16.20	15	18.7670	5.4270	0.1059	12.5280	3.2510	0.1704	16.9160	7.8420	0.4939
16.30	1	1.0000	0.0000	1.0000	1.0000	0.0000	1.0000	0.9634	0.0353	0.8454
18.00	36	44.3620	37.5380	0.1723	40.2530	26.9660	0.4128	42.2930	40.1110	0.3204

19.11	15	15.0000	0.0000	1.0000	8.3210	2.1770	0.0000	12.1020	4.6580	0.1795
19.12	15	15.0000	0.0000	1.0000	11.4640	1.1400	0.0009	14.9870	0.0135	0.9136
20.00	15	15.0000	0.0000	1.0000	12.8330	3.1760	0.2240	13.1390	3.6450	0.3296
23.00	6	6.0000	0.0000	1.0000	5.6400	0.3587	0.5624	5.8860	0.1538	0.7711
24.00	3	3.6880	5.2990	0.7649	4.3020	2.5720	0.4168	4.4900	5.1470	0.5112
25.11	18	16.9380	29.7500	0.8456	14.0890	12.6490	0.2715	13.4320	24.6830	0.3579
25.12	18	16.9380	29.7500	0.8456	13.4240	13.7380	0.2170	12.8860	23.1020	0.2873
26.00	3	3.0000	0.0000	1.0000	3.0000	0.0000	0.9999	3.0000	0.0000	1.0000
27.00	55	55.0000	0.0000	1.0000	50.7910	5.3290	0.0683	54.9570	0.0445	0.8401
28.11	256	295.4100	776.0800	0.1572	170.9400	205.1800	0.0000	213.0300	761.2700	0.1193
28.12	256	296.6100	760.4400	0.1408	172.7800	202.0300	0.0000	216.0100	758.7900	0.1466
28.20	168	208.4800	1533.1000	0.3012	125.6300	244.3700	0.0067	143.2300	786.3600	0.3771
28.30	165	172.6100	1522.0000	0.8454	76.0230	162.1600	0.0000	78.1800	383.9900	0.0000
29.00	4	5.2610	6.5390	0.6218	5.0790	3.1480	0.5430	5.0670	5.4600	0.6478

i Model 2 in section 4.3.1.
j Model 4 in section 4.3.1.
k Model 6 in section 4.3.1.

model, the probability P of a deviation from the mean at least as large as that observed is also given for each food web version. When the model necessarily matches the observed E perfectly, with 0 predicted variance (for example, model 1 applied to food web 26), P has been set to 1 to signify that no better fit is possible. Figure 12 gives the frequency histograms of P for each of the six models. If all 45 food web versions were randomly sampled from a universe of food webs in which one of the models

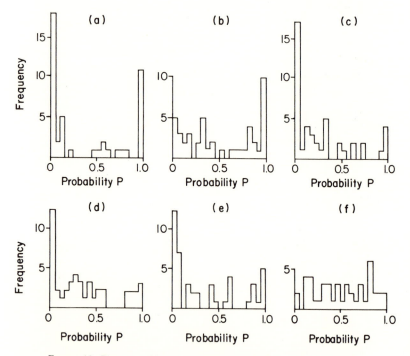

FIGURE 12. Frequency histograms of the probabilities, under a normal approximation, of standardized deviations as large as or larger than those observed between the number of edges in the niche overlap matrix and the theoretical expectation, for all food web versions. (a) Model 1. (b) Model 2. (c) Model 3. (4) Model 4. (e) Model 5. (f) Model 6. Models 1, 3, 4, and 5 describe either the mean or variance of observed niche overlaps poorly. (Based on Table 8.)

were true, the frequency histogram of P values for that model would approximate a uniform distribution (a horizontal straight line). The assumption of random sampling of food web versions is clearly invalid here. Nevertheless when the distribution of P is heavily concentrated in the interval from 0 to 0.05, the underlying model provides a poor description of either the mean or variation of overlaps in these food webs. Figures 12a,c,d,e suggest that models 1, 3, 4, and 5 describe either the mean or variation of observed niche overlaps poorly.

A plot of the observed overlap on the ordinate against the predicted mean on the abscissa should approximate a straight line of slope 1. Figure 13 shows that such a plot is approximately linear for model 2 (linear correlation coefficient = 0.97), but the slope is systematically too low. In a least-squares regression, the slope is 0.58 with sample standard deviation of 0.02. Thus the means predicted by model 2 are systematically too large,

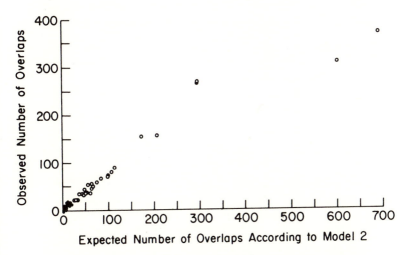

FIGURE 13. The observed number of edges in the niche overlap graph, or the number of overlaps among the diets of the predators, as a function of the theoretical expectation of model 2, for all food web versions. The relation is nearly linear, but the slope is smaller than 1. (Based on Table 8.)

when compared with the observed overlaps. The same holds for model 1. The means predicted by models 3 and 4 are systematically too small, though the linearity between observed and predicted means is excellent for all four models.

The same plot for model 6 (Figure 14) is again approximately linear (linear correlation coefficient = 0.98). The regression slope of 1.01 with standard deviation 0.03 is not distinguishable from 1. The y-intercept of the regression line does not differ noticeably from 0. The predator-prey relations assumed in model 6 lead to an excellent description of niche overlap.

The success of model 6 is compatible with the success of the models assuming truncated binomial row sums and truncated binomial column sums in section 4.2.1. Under model 6, the distribution of row sums in a food web matrix will be binomial, and so also will be the distribution of column sums. If the probability of 0 in these distributions is not large, they will not differ detectably from 0-truncated binomial distributions.

Model 5 differs from model 6 only in assuming that the number A of 1's in the food web matrix is fixed, rather than

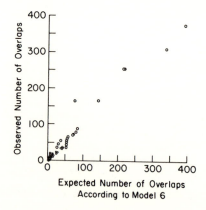

FIGURE 14. The observed number of edges in the niche overlap graph, or the number of overlaps among the diets of the predators, as a function of the theoretical expectation of model 6, for all food web versions. The relation is nearly linear and the slope is close to 1. (Based on Table 8.)

68

binomially distributed. Conditional on A, the two models are identical. One might suspect, therefore, that the two models would predict comparable mean values for E. Model 6, assuming binomially distributed A, predicts a higher variance in E and hence is more tolerant than model 5 of deviations between the observed E and the predicted mean. Comparison of the means predicted by models 5 and 6 in Table 8 shows that for each food web version the two are close. Figure 15 confirms the excellent fit of model 5's predicted means to the observed E. The regression slope is again 1.01 with standard deviation 0.03, and the linear correlation coefficient is 0.98.

The closer approximation to a uniform distribution of the frequency histogram of P in Figure 12f compared with that in Figure 12e thus shows that the variation of the observed E about a mean common to models 5 and 6 is more plausibly described by the assumption of a binomially distributed A than by the assumption that A is fixed. It could also be argued that the assumption of a fixed A is less plausible a priori.

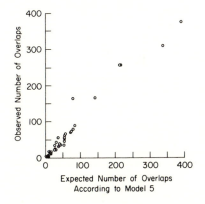

FIGURE 15. The observed number of edges in the niche overlap graph, or the number of overlaps among the diets of predators, as a function of the theoretical expectation of model 5, for all food web versions. The relation is nearly linear and the slope is close to 1. (Based on Table 8.)

69

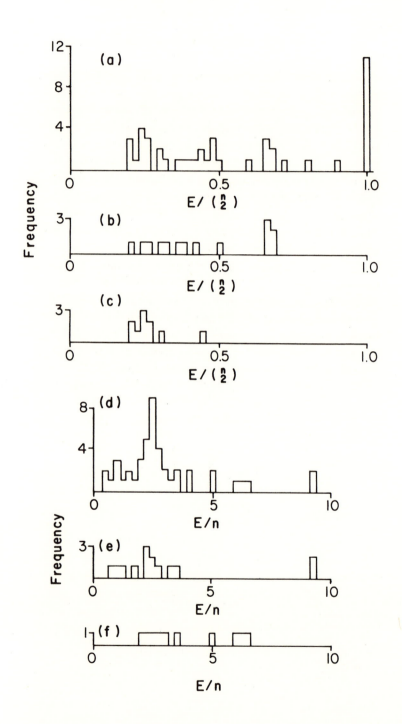

The net result of this subsection is to provide a simple random model for the predator-prey relations in food webs. The model (model 6) is that every kind of predator in a given food web has a constant and independent probability of preying on every kind of prey in that food web. This model has two desirable features. First, it has only one parameter, which is readily estimated given any real food web. Second, it approximates reasonably well the mean and variance of the number of trophic niche overlaps among the kinds of predators in the food web. The model was developed in order to define a realistic universe of random food webs that would in turn define a realistic universe of random niche overlap graphs, in order to compare the frequency of interval food webs among observed and random food webs. Because of its extreme, indeed scarcely credible, simplicity, the model is interesting in itself as a description at the level of an ecological community of an ensemble of interactions the details of which are known to be complicated. Though such a study would be a digression here, it would seem worth investigating further whether there are other quantitative characteristics of a food web in addition to niche overlaps that are approximately described by this model.

4.3.2 *The Distribution of Niche Overlaps*

There are at least two natural scales on which to examine the frequency distribution among food webs of the number E of

---

FIGURE 16. (a),(b),(c) Frequency histograms of $E \left/ \binom{n}{2} \right.$, where E is the number of edges in the niche overlap graph and n is the number of kinds of predators; $E \left/ \binom{n}{2} \right.$ is the fraction of possible niche overlaps which actually occur. (a) All food web versions. (b) Interval community food web versions. (c) Non-interval community food web versions. (d),(e),(f) Frequency histograms of E/n, the number of niche overlaps per kind of predator. (d) All food web versions. (e) Interval community food web versions. (f) Non-interval community food web versions. (Based on Table 5.)

niche overlaps among the n kinds of predators in a food web. Since the maximum possible value of E is $n(n - 1)/2$, the ratio $2E/(n(n - 1))$ relates observed niche overlap to the maximum possibilities created by the number of predators in the food web. The ratio E/n is the average number of niche overlaps per predator in each food web.

Figure 16a is a frequency histogram of the observed number E of niche overlaps as a fraction of the maximum number $n(n - 1)/2$ for all 45 food web versions. The peak at the right end of the histogram, where $E = n(n - 1)/2$, is created by the appearance in sink food webs of sets of predators, all of which have overlapping niches; the histograms of community food web versions (Figures 16b,c) show no such peak. The interval food web versions (Figure 16b) appear to have more niche overlap than do the non-interval food web versions (Figure 16c). None of the community food webs has values of $2E/(n(n - 1))$ lying below 0.2 or above 0.8, those regions where interval graphs are most likely to arise by chance (Figure 17 below).

The ratio E/n of niche overlaps per predator ranged from 0.5 to 9.14 among the 45 food web versions, with a mean of 2.88. The apparent peak for all 45 food web versions in the frequency histogram of E/n (Figure 16d) is the rescaled peak from the extreme right of Figure 16a. It disappears when only the 24 community food web versions are considered (Figures 16e,f). On this scale it is difficult to argue for a difference between interval and non-interval food webs. Again there is no pronounced tendency toward high or low extremes. If we make allowance for the dependence among multiple versions of a single food web, and assume no bias in the sampling of food webs according to their values of E/n, the observations in Figures 16e,f are probably insufficient to reject the model that among community food webs E/n is uniformly distributed over some finite range falling within the interval from 0 to 10. If anything, the histograms in Figures 16b,c,e,f suggest that the niche overlaps E are less likely than one might expect from a uniform distribution to be either extremely high or extremely low, on either scale.

4.4 SUMMARY

In order to know whether it is surprising that the frequency of interval food webs is as high as the frequency observed in Chapter 3, it is necessary first to estimate a universe of possible food webs from which the observed food webs are drawn, and second to estimate the frequency of interval food webs in that universe. This chapter is devoted to the first task, a statistical analysis of the 45 observed food web versions. Chapter 5 is devoted to the second task.

The variation in the number of kinds of prey taken by each kind of predator in an interval community food web can be fairly well described by assuming that every predator chooses to prey or not to prey on every kind of available prey independently and with a constant probability. This simple model is not adequate for interval sink food webs or for non-interval food webs.

Similarly, in an interval community food web, the variation in the number of different kinds of predators taking each kind of prey can be fairly well described by assuming that each kind of prey is exposed to a risk of predation that is independent and constant for every predator in the community. This simple model is not adequate for interval sink food webs or for non-interval food webs.

When compared with predators in non-interval community food webs, predators in interval community food webs maintain roughly the same average number of kinds of prey taken, although fewer kinds of prey are available to them, by taking a higher proportion of the kinds of available prey. This higher proportion reduces the variation in the number of prey taken among predators within a community.

Among community food webs, the ratio of the number of kinds of prey to the number of kinds of predators closely approximates a constant near 3/4. Among sink food webs, the ratio, also only slightly variable, is higher because a sink food web includes all the prey of the chosen sinks, without necessarily including all the other predators of those prey. The ratio of 3/4 can be predicted as the ratio of the mean, over all community food webs, of the average number of kinds of prey per predator in a community, to the mean of the average number of kinds of predators per prey in a community. This invariance in the ratio of kinds of prey to kinds of predators predicts correctly the covariation among community food webs of the average number of kinds of prey per predator with the average number of kinds of predators per prey.

The observed distributions of the lengths of food chains in four food webs with many kinds of predators are roughly consistent with Hutchinson's (1959) predictions for Eltonian predator chains, but the quantitative details remain to be explained.

Both the mean and the variation from the mean of the number of niche overlaps are very well described by assuming that every predator in a given food web has a constant and independent probability of preying on every prey in that food web.

As a proportion of the maximum possible number of niche overlaps, the distribution among the observed food webs of the number of niche overlaps is peaked at the upper extreme for sink food webs. Among community food webs, those which are interval appear to have more niche overlap than do those that are non-interval.

As a multiple of the number of predators in the food web, the distribution of niche overlaps is roughly uniform in community food webs, with little difference between those that are interval and those that are non-interval.

CHAPTER FIVE

Frequencies of Interval Food Webs in Observed and Monte Carlo Samples

First it is fundamental to realize
No two of anything may be alike.
That dawn out there that paints
   those loitering skies
Around St. Ceil's pale lemon, and
   tints white
Pilasters on its spire the tastiest
   lime,
Cannot come up the same another
   time . . .

L. E. Sissman (1967)

5.1 MATCHING FOOD WEB CHARACTERISTICS

The analytical problem of enumerating interval graphs, whether exactly or asymptotically, has been posed to combinatorialists and graph theorists for several years (Harary and Palmer, 1973, p. 265). Although asymptotic results, for very large numbers of vertices, may soon be available, we must resort for most of our estimates to Monte Carlo simulation and sampling.

For each version of each food web with four or more kinds of predators, and for each of the six models of the food web matrix presented in section 4.3.1, one hundred artificial food web matrices having the characteristics assumed by the model were generated using a pseudo-random number generator. Specifically, for model 1, the entries in each row of the real food web matrix were randomly permuted. For model 2, each entry of a given row was set to 1 with a probability equal to the p-parameter of a matching 0-truncated binomial distribution (section 4.2.1). If the artificial row thus generated

75

had no 1's, it was rejected and another try was made. For models 3 and 4, the same procedures for models 1 and 2, respectively, were applied to columns instead of rows. For model 5, all entries in the real food web matrix were randomly permuted. For model 6, each entry was set to 1 with probability $p = A/(mn)$ and to 0 with probability $1 - p$.

Each artificial food web was tested for being interval or non-interval using the algorithm previously applied to the real food webs (Chapter 3). Table 9 shows the fraction of artificial food webs that were interval in 100 trials. Since food webs with two or three predators are necessarily interval, the corresponding frequencies are reported as 1.00. These food webs have no effect on the conclusions that will be drawn.

Let f_{ij} be the fraction that are interval of the artificial food webs having the parameters of real food web version i generated according to model j. For example, for food web version $i = 1.22$ and model $j = 1$, $f_{ij} = 0.36$, according to Table 9. We shall assume that f_{ij} represents the true proportion of all food webs that are interval in the jth model universe with the parameters specified by real food web i. We neglect the likelihood that f_{ij} differs from the true proportion, say p_{ij}, because we have sampled only 100 artificial food webs from this universe. The standard deviation of f_{ij} in a random sample of size 100 is $[p_{ij}(1 - p_{ij})/100]^{1/2}$, which cannot exceed 0.05. At the cost of more computing time, additional artificial food webs could be generated. But the clarity of the conclusions that follow makes it unlikely that a larger artificial sample size would change the results.

Let S be a set of real food web versions, such as the set of community food web versions. If the ith food web version in the set has probability f_{ij} of being interval according to model j, then the expected number of interval food webs in the set S, according to model j, is

$$\mu_{Sj} = \Sigma_{i \in S} f_{ij}. \tag{5.1}$$

TABLE 9. The estimated fraction of food webs which are interval according to seven models, given the characteristics of each food web version.

Code[a] Number	Model Number						
	1[b]	2	3	4	5	6	7[c]
1.11	0.29	0.45	0.04	0.22	0.07	0.15	0.0145
1.12	0.46	0.51	0.01	0.13	0.05	0.05	0.0000
1.21	0.66	0.86	0.88	0.95	0.77	0.77	0.4100
1.22	0.36	0.71	0.66	0.88	0.73	0.70	0.1800
1.30	0.04	0.26	0.00	0.08	0.00	0.03	0.0890
1.41	0.00	0.01	0.00	0.00	0.00	0.00	0.1280
1.42	0.00	0.00	0.00	0.00	0.00	0.00	0.0990
2.11	0.10	0.34	0.08	0.19	0.17	0.19	0.0000
2.12	0.00	0.01	0.00	0.01	0.00	0.00	0.0546
4.00	1.00	1.00	1.00	1.00	1.00	1.00	1.0000
5.11	1.00	1.00	0.97	1.00	0.97	0.97	1.0000
5.12	0.00	0.00	0.00	0.00	0.00	0.00	0.0000
5.13	0.00	0.01	0.00	0.00	0.00	0.00	0.0000
7.11	0.00	0.00	0.00	0.00	0.00	0.00	0.0808
7.12	0.00	0.00	0.00	0.00	0.00	0.00	0.0959
8.10	0.17	0.49	0.65	0.64	0.29	0.22	0.0100
8.20	0.01	0.09	0.01	0.02	0.00	0.00	0.0000
8.30	0.00	0.06	0.01	0.02	0.00	0.00	0.0000
10.11	1.00	1.00	0.98	0.99	0.99	1.00	1.0000
10.12	1.00	1.00	0.95	1.00	1.00	0.99	1.0000
11.11	0.00	0.07	0.12	0.21	0.00	0.00	0.1479
11.12	0.00	0.01	0.02	0.12	0.00	0.00	0.0978
12.00	0.83	0.91	0.79	0.80	0.77	0.72	0.1800
13.00	1.00	1.00	1.00	1.00	0.82	0.74	0.0000
15.11	0.00	0.00	0.00	0.00	0.00	0.00	0.2117
15.12	0.00	0.00	0.00	0.00	0.00	0.00	0.0962
15.13	0.00	0.00	0.00	0.00	0.00	0.00	0.0821
16.10	1.00	1.00	1.00	1.00	1.00	1.00	1.0000
16.20	0.89	0.95	1.00	0.99	0.84	0.79	0.3000
16.30	1.00	1.00	1.00	1.00	1.00	1.00	1.0000
18.00	0.16	0.50	0.21	0.27	0.34	0.32	0.0027
19.11	1.00	1.00	1.00	1.00	0.86	0.85	1.0000
19.12	1.00	1.00	1.00	1.00	1.00	1.00	1.0000
20.00	1.00	1.00	0.99	1.00	0.98	0.96	1.0000
23.00	1.00	1.00	1.00	1.00	1.00	1.00	1.0000
24.00	1.00	1.00	1.00	1.00	0.99	0.98	1.0000
25.11	0.95	0.98	0.64	0.86	0.81	0.83	0.0500
25.12	0.99	0.99	0.54	0.78	0.85	0.81	0.0500
26.00	1.00	1.00	1.00	1.00	1.00	1.00	1.0000
27.00	1.00	1.00	1.00	0.95	1.00	0.99	1.0000
28.11	0.00	0.01	0.00	0.00	0.00	0.00	0.0039
28.12	0.00	0.02	0.00	0.00	0.00	0.00	0.0039
28.20	0.00	0.00	0.00	0.00	0.00	0.00	0.0553
28.30	0.00	0.08	0.00	0.00	0.00	0.00	0.0000
29.00	1.00	1.00	1.00	1.00	1.00	0.98	0.9286

[a] Code number of food web version, as in Table 5.

[b] For models 1 to 6, defined in section 4.3.1, for each food web version, the fraction given is the proportion of 100 pseudo-random food webs that were interval. The fraction is taken as 1 for food webs with less than 4 predators.

[c] For model 7, defined in section 5.2, for each food web version, the fraction of food webs which are interval is exact wherever the analytical results in section 5.2 can be used, is otherwise based on 100 pseudo-random food webs when there are 10 or fewer predators in the food web version, and when there are more than 10 predators is based on interpolation from the results with 10 predators.

If in addition we assume that each food web version is sampled from the jth model universe independently of every other food web version in the set S, then the variance in the number of interval food web versions in the set S, according to model j, is

$$\sigma^2_{Sj} = \Sigma_{i\in S} f_{ij}(1 - f_{ij}). \qquad (5.2)$$

When the number of food web versions in the set S is reasonably large, the distribution of the number that are interval can be taken as approximately normal with the mean μ_{Sj} and the variance σ^2_{Sj}. Hence the quantity

$$z_{Sj} = \text{(observed number of interval food web}$$
$$\text{versions in } S - \mu_{Sj})/\sigma_{Sj} \qquad (5.3)$$

should be normally distributed with mean 0 and standard deviation 1 if the food web versions are randomly sampled from the jth model universe.

Table 10 gives the observed number of interval food web versions, the mean μ_{S6}, the standard deviation σ_{S6}, and the normal variate z_{S6}, for 12 different sets S of food web versions for model $j = 6$ only. The statistics presented in Table 10 for model 6 have also been calculated for the other five models and lead to exactly the same qualitative conclusions with only minor quantitative differences. Since model 6 describes actual niche overlaps (section 4.3.2), it seems sufficient to present the results for model 6 alone.

The least plausible step in the argument that the statistic z_{Sj} defined by equation (5.3) has the standardized normal distribution is the assumption that all 45 food web versions listed in Table 5 are independently sampled. Though different food webs may be independent, different versions of a single food web can hardly be independent. In Table 10, "version A" food webs are all the food webs obtained by counting only entries of $+1$ as 1 and setting doubtful entries of -1 and -2 to 0. The code numbers in Table 5 of version A food webs have either 0 or 1 in the second place to the right of the decimal

78

TABLE 10. The observed numbers of interval food web versions in 12 sets of food web versions, the corresponding means and standard deviations of the number of interval food webs according to model 6, and one-tailed tests for the excess of the observed over the expected frequency of interval food webs.

Set of food web versions	Versions in set	Observed number interval	μ[a]	σ[b]	z[c]
All food web versions	45	33	20.04	1.50	8.65
Interval food web versions	33	33	20.04	1.50	8.65
Community food web versions	24	14	4.83	1.13	8.11
Sink food web versions	20	18	14.47	0.88	4.01
All version A[d] food webs	31	24	16.49	1.35	5.56
Interval version A food webs	24	24	16.49	1.35	5.56
Community version A food webs	14	9	3.27	0.93	6.16
Sink version A food webs	16	14	12.48	0.87	1.74
All version B[e] food webs	31	23	15.28	1.22	6.33
Interval version B food webs	23	23	15.28	1.22	6.33
Community version B food webs	14	8	2.89	0.82	6.21
Sink version B food webs	16	14	11.65	0.79	2.99

[a] Expected number of interval food web versions in the set specified by each row, calculated from eq. (5.1) using $j = 6$ and f_{ij} from Table 9.

[b] Standard deviation of the number of interval food web versions in the set specified by each row, calculated as the square root of eq. (5.2) using $j = 6$ and f_{ij} from Table 9.

[c] Standardized difference of the observed from the expected number of interval food web versions in the set specified by each row, calculated according to eq. (5.3). Assuming that the observed number of food web versions is approximately normally distributed with the calculated mean and standard deviation, the probability of a larger excess than $z = 3.1$ is less than 0.001 (one-tailed test).

[d] Version A food webs have code numbers in Table 5 with 0 or 1 in the second place to the right of the decimal point.

[e] Version B food webs have code numbers in Table 5 with 0 or 2 in the second place to right of the decimal point.

point. "Version B" food webs are those obtained by setting entries of -1 and $+1$ to 1. They have code numbers in Table 5 with either 0 or 2 in the second place to the right of the decimal point. Among version A or version B food webs, the assumption of independence is much more plausible than among all food web versions.

Since a standardized normal variate exceeds the value 3.1 with a probability of 0.001, we adopt 3.1 as a conservative critical value for z_{S6}.

It is then evident from Table 10 that there are more interval food webs in the set of all version A food webs, in the set of interval version A food webs, and in the set of community version A food webs, than could likely arise by random sampling from the regions specified by the observed food webs in model universe 6. The frequency of sink version A food webs does not significantly exceed the frequency expected from random samples of the corresponding regions of model universe 6. The version B food webs lead to exactly the same conclusion.

The same relative pattern of z_{S6} values is evident among the food web versions when different versions of a single food web are not excluded. The apparent statistical significance of $z_{S6} = 4.01$ for the set S of sink food web versions results from the underestimation of the corresponding variance, due to the omission of terms describing the positive covariance between different versions of the same food web.

In summary, whether the food webs analyzed are based on information considered certain (version A) or include also some inferential information (version B), substantially and significantly more of them are interval than expected from a random sample of artificial food webs having the same average number of randomly placed directed edges (model universe 6). That community food webs are interval far more often than chance alone would provide is a striking finding that needs explanation.

The number of sink food webs that are interval is not significantly greater than the number to be expected from random sampling of the corresponding regions of model universe 6. For either version A or version B, 14 of the 16 distinct sink food webs in Table 5 are interval. The two exceptions, food webs 8.2 and 8.3 of Kohn (1959), are discussed in detail in section 3.2. One of them describes a composite of two different habitats; the other, based on a single habitat, becomes interval when diets based on very few observations are excluded. Thus all of the sink food webs are consistent with a one-dimensional niche space in single habitats. We therefore interpret the statistical insignificance of z_{S6} for sink food webs as showing that the collection of sink food webs defined by model 6 shares with the real sink food webs a very high proportion of interval food webs, just as the London suburbs share with London the practice of driving on the left. This high frequency of interval sink food webs may well be due to investigators choosing sets of sinks with "interesting" overlapping (food) niches.

5.2 MATCHING NICHE OVERLAP

We now define another model of food webs and compare the observed frequency of interval food webs with the approximate expected frequency.

A classical definition of a random (undirected) graph that has n labeled vertices (kinds of predators) and E (undirected) edges is that the E edges are chosen at random among the $n(n-1)/2$ possible edges, so that all $\binom{n(n-1)/2}{E}$ possible graphs are equiprobable (Erdös and Rényi, 1960, p. 17).

Model 7: Random niche overlap graph. If a given food web version has n kinds of predators and E niche overlaps, suppose that its niche overlap graph is a random graph (in the sense of Erdös and Rényi) on n vertices with E edges.

For $n = 10$ and $n = 40$ vertices and for several values of E, Table 11 shows the number of pseudo-random graphs that were interval in 100 Monte Carlo simulations. All graphs with 0, 1, 2, 3, $\binom{n}{2} - 1$, and $\binom{n}{2}$ edges are necessarily interval. For $n > 3$, Dr. Thomas Mueller has calculated some of the probabilities that a random graph with E edges is interval:

$$1 - 48/[(n^2 - n - 4)(n + 2)(n + 1)] \qquad \text{if } E = 4;$$

$$1 - \frac{240(n-4)(n+(23/5))}{[(n^2-n-4)(n+2)(n+1)(n^2-n-8)]} \qquad \text{if } E = 5;$$

$$(32n - 88)/[(n + 1)(n^2 - n - 4)] \qquad \text{if } E = \binom{n}{2} - 3;$$

$$4/(n + 1) \qquad \text{if } E = \binom{n}{2} - 2.$$

$$(5.4)$$

TABLE 11. The number of pseudo-random undirected graphs on 10 and on 40 vertices which were interval in 100 Monte Carlo trials, and some exact expected frequencies of interval graphs.

$n = 10$ vertices			$n = 40$ vertices		
Edges[a]	Frequency[b]	Expected[c]	Edges	Frequency	Expected
0	100		0	100	
3	100		3	100	
6	86		17	39	
7	51		20	22	
8	23		25	0	
9	11		156	0	
18	0		312	0	
27	0		468	0	
36	1		624	0	
42	36	24.5	777	3	1.87
43	44	36.4	778	25	9.76
44	100		779	100	
45	100		780	100	

[a] Number of undirected edges in each set of 100 pseudo-random graphs.

[b] Number of interval pseudo-random graphs in the set with the given number of edges.

[c] Expected number of interval graphs in a set of 100 random graphs with the given number of edges, calculated according to eqs. (5.4).

The expected frequencies in 100 trials, based on the last two of equations (5.4), are also shown in Table 11. We discuss the discrepancies between the Monte Carlo and the expected frequencies in section 5.3.

The Monte Carlo frequencies for $n = 10$ are graphed in Figure 17a and for $n = 40$ in Figure 17b. The horizontal axis of each graph is labeled both by the number of edges as a fraction of $\binom{n}{2}$ and by the number of edges as a multiple of n. The purpose of the figures is to see if there is some obvious scaling of the number of edges in relation to the number of vertices such that the probability of an interval graph then becomes an invariant, or only slowly varying, function of n.

Though neither rescaling is perfect, rescaling E by $\binom{n}{2}$ is

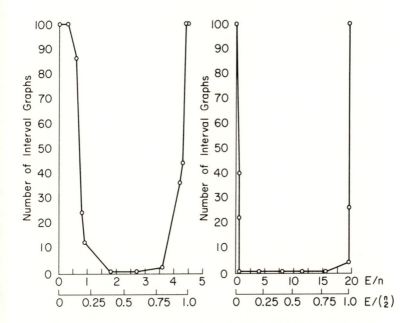

FIGURE 17. The number of pseudo-random undirected graphs, out of 100, which were interval as a function of E/n (upper abscissa) or as a function of $E/\binom{n}{2}$ (lower abscissa). n is the number of vertices and E is the number of undirected edges. (a, left) $n = 10$. (b, right) $n = 40$. (Based on Table 11.)

much better than rescaling by n, for the two values of n used. Over the broad range of $E/\binom{n}{2}$ from 0.25 to 0.75 the probability of an interval graph is 0 or close to 0 for both values of n. For values of $E/\binom{n}{2}$ approaching 0 from above or 1 from below, the probability of an interval graph ascends steeply from 0 to 1 for both values of n. On the other hand, when $n = 10$, the probability of an interval graph approaches 1 when E/n approaches 4.5 from below, while when $n = 40$, the probability of an interval graph is 0 for E/n ranging from 0.625 to at least 15.

83

For any value of $E / \binom{n}{2}$, the estimated probability of an interval graph when $n = 10$ is greater than or equal to the estimated probability when $n = 40$. We shall act on the conjecture that, for a given value of $E / \binom{n}{2}$, the probability that a random graph is an interval graph is a non-increasing function of n. Hence, when n is larger than 10, using the probability of an interval graph for $n = 10$ will yield an overestimate, and therefore minimize the difference between the calculated expected frequency of interval graphs and any unusually high observed frequency. Approximating the probability of an interval graph for n larger than 10 by the estimated probability for $n = 10$ also overstates the variance in the number of interval graphs. The reason is that all the estimated values of f_{i7} are less than 0.5. The closer f_{i7} is to 0.5, the larger is the variance term $f_{i7}(1 - f_{i7})$. Thus the approximation decreases the chance of attributing statistical significance to a high observed frequency of interval graphs.

For each combination of n vertices (kinds of predators) and E edges in the niche overlap graphs in Table 5, the probability of an interval graph according to model 7 was estimated (Table 9). Exact probabilities were calculated using equations (5.4) and the preceding comments wherever possible. For each of the remaining food web versions with nine or fewer predators, 100 random niche overlap graphs were artificially simulated using exactly the observed number of edges in the real niche overlap graph. For each of the remaining food webs with 10 or more predators, an overestimate of the probability of an interval graph was obtained by linear interpolation for the observed value of $E / \binom{n}{2}$ in the probabilities for $n = 10$ in Table 11. Exact probabilities from equations (5.4) were used wherever possible in forming the interpolation table from Table 11. Since the median number of predators in the 45 food web

84

FREQUENCIES OF INTERVAL FOOD WEBS

TABLE 12. The observed numbers of interval food web versions in 12 sets of food web versions, the corresponding means and standard deviations of the number of interval food webs according to model 7, and one-tailed tests for the excess of the observed over the expected frequency of interval food webs.

Set of food web versions	Versions in set	Observed number interval	μ^a	σ^b	z^c
All food web versions	45	33	16.37	1.43	11.66
Interval food web versions	33	33	15.43	1.10	16.03
Community food web versions	24	14	2.95	1.26	8.73
Sink food web versions	20	18	13.42	0.66	6.96
All version A food webs	31	24	13.61	1.16	8.92
Interval version A food webs	24	24	13.14	0.98	11.14
Community version A food webs	14	9	2.19	0.96	7.09
Sink version A food webs	16	14	11.42	0.66	3.92
All version B food webs	31	23	12.24	1.08	9.94
Interval version B food webs	23	23	11.80	0.88	12.78
Community version B food webs	14	8	1.82	0.86	7.19
Sink version B food webs	16	14	10.42	0.66	5.44

[a] Expected number of interval food web versions in the set specified by each row, calculated from eq. (5.1) using $j = 7$ and f_{ij} from Table 9.

[b] Standard deviation of the number of interval food web versions in the set specified by each row, calculated as the square root of eq. (5.2) using $j = 7$ and f_{ij} from Table 9.

[c] See footnote c of Table 10.

versions is 11, this procedure overestimates the probability of an interval graph for about half the food web versions.

From the last column of Table 9, the expected number μ_{S7} of interval food web versions in the set S, the variance σ_{S7}^2 in the number of interval food web versions in S, and the standardized normal deviate z_{S7} were calculated (Table 12) according to equations (5.1) to (5.3), for 12 sets S of real food web versions.

The conclusion can be guessed by comparing Figures 16 and 17: most of the observed values of $E \Big/ \binom{n}{2}$ fall in the region where the probability of an interval graph is close to 0. All 12 values of z_{S7} are so much larger than the critical value 3.1 that the high frequency of interval graphs among the observed niche overlap graphs is extremely unlikely to have arisen randomly according to model 7.

5.3 CALIBRATING THE LOTTERY

Exact values rather than Monte Carlo values from Table 11 were used wherever possible in estimating the mean and variance of the number of interval food webs expected in Table 12. Still, the discrepancies between the Monte Carlo frequencies and the exact expected frequencies in Table 11 are disturbing. In all four cases where the comparison has been made, the Monte Carlo frequencies exceed the exact. In the worst case, where $n = 40$ and $E = 778$, the discrepancy between the exact expectation of 9.76 and Monte Carlo frequency of 25 is improbably large, by any reasonable criterion. That the Monte Carlo frequencies overstate the exact expectations is less cause for concern than if they understated them, since the observed frequency of real interval food webs is still significantly large. But discrepancies of such magnitude raise the specter of undetected error or systematic bias.

86

To investigate the worst case, where $n = 40$ and $E = 778$, another 100 artificial niche overlap graphs were generated by the algorithm used previously, starting with a different seed for the pseudo-random number generator. This generator is named "deal" in the language called APL (Gilman and Rose, 1974). A pseudo-random sample without replacement of 778 positive integers less than or equal to 780 was used to specify the 778 edges *present* in each artificial niche overlap graph. The two edges missing from the complete graph on 40 vertices were printed out for each of the 100 graphs. The machine implementation of the Fulkerson-Gross (1965) algorithm reported that of these 100 graphs, 21 were interval. All 100 were verified by hand. However, certain edges appeared to be absent from the graphs with far more than expected frequency.

Using the same pseudo-random number generator, another 100 artificial niche overlap graphs were then generated. This time, however, a pseudo-random sample without replacement of two positive integers less than or equal to 780 was used to specify the edges *absent* from the complete graph on 40 vertices. All 100 graphs were again verified by hand against the machine version of the Fulkerson-Gross (1965) algorithm. Again no discrepancies were found. Of these 100 graphs, only four were interval.

These results suggest that the discrepancies in Table 11 are due, not to failures of the algorithm for checking whether a graph is interval, but to inadequacies of the procedure for generating pseudo-random graphs. No errors could be detected in the program for specifying a graph given the list of edges present in it.

Investigation of the pseudo-random number generator soon showed systematic bias. Ten thousand samples without replacement of two positive integers less than or equal to 3 were tallied according to which integer appeared first and which integer appeared second (Table 13). The expected frequency with

TABLE 13. Frequency distributions of the integers occurring in two sets of 10,000 pseudo-random samples without replacement from the integers 1, 2, 3.

	First set of 10,000 samples[a]			
	1	2	3	X^{2b}
First number	2228[c]	4418	3354	719.61
Second number	3261	3431	3308	4.62

	Second set of 10,000 samples[d]			
	1	2	3	X^2
First number	2254	4557	3189	804.95
Second number	3277	3277	3446	5.71

[a] Each sample contained 2 positive integers drawn without replacement from the integers 1, 2, 3; thus the first number of a typical sample might be 2 and the second number 3.

[b] Chi-squared statistic with 3 df for goodness-of-fit to the uniform distribution, with expectation $10,000/3 = 3333.3$; calculated separately for the first number and for the second number, within each set of 10,000 samples. If the samples were truly random, the probability of a value of X^2 exceeding 16.27 would be less than 0.001.

[c] Frequency of 1 as the first number in 10,000 samples.

[d] The seed ("random link" in APL) for the first 10,000 samples was 482613421; for the second, 772695759.

which each integer 1, 2, or 3 should appear either first or second in the sample is $10,000/3 = 3333.3$. The integers listed as second in the samples were distributed about as expected ($X^2 = 4.62$ with 3 df). But there were far fewer 1's and far more 2's than expected among the first numbers in the samples ($X^2 = 719.61$ with 3 df). Another 10,000 samples starting from a different seed for the pseudo-random number generator revealed an identical pattern (Table 13). Many other tests not described here confirmed this pattern.

To see how much damage this systematic bias had inflicted on the other Monte Carlo results, the number of niche overlaps was recorded for each of 10,200 pseudo-random food webs (defined in section 5.1). These 10,200 pseudo-random food webs consist of 100 for each one of 17 arbitrarily selected real food web versions and for each one of the six food web models described in section 4.3.1; 10,200 = 100 × 17 × 6. From each set of 100, the sample mean and sample variance of niche overlaps were calculated.

Each sample variance, divided by the theoretical variance of niche overlap (Table 8) for the same food web version and model, and multiplied by $100 - 1 = 99$, should then have the distribution of χ^2 with 99 *df* if the 100 observations were randomly sampled from a population with the theoretical variance (Snedecor and Cochran, 1967, p. 76). The statistic 99(sample variance)/(theoretical variance) was transformed (Snedecor and Cochran, p. 233) to what should be a standardized normal variate if the pseudo-random number generator were perfect, the numerical values of the theoretical variances were correct, and the several normal approximations used were exact.

Figure 18 is a frequency histogram of the resulting $102 = 17 \times 6$ values. The five observations on the far left are cases

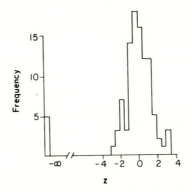

FIGURE 18. A frequency histogram of the transform z of a statistic that compares the sample variance of the number of niche overlaps in 100 pseudo-random food webs with the theoretical variance. The transformed values number $102 = 17$ food web versions $\times$ 6 models. If the pseudo-random number generator were random, the theoretical variances were correct, and the normal approximations exact, this histogram, excepting the 5 cases with 0 variance on the left, would approximate a normal distribution with mean 0 and variance 1. (Based in part on Table 8.)

where the combination of food web version and model specify exactly the number of niche overlaps, with no sampling variance. The remaining 97 values in the bell-shaped region on the right of Figure 18 have a sample mean of 0.1253 and sample variance of 1.4614. Using exactly the same test as above for comparing 1.4614 with the theoretical variance of 1 gives the probability of a larger deviation by chance as 0.002. This probability is not lower than the significance level adopted for comparing the observed frequency of interval food webs with the frequencies expected from the various models.

If one takes the variance of the 97 values as 1, then the sample mean of 0.1253 does not differ significantly from 0 at the 0.001 level. Thus the sample variances of niche overlap in the Monte Carlo simulations roughly approximate the theoretical variances in Table 8.

The difference between the 102 sample means of niche overlap and the corresponding 102 theoretical means of niche overlap, divided by the theoretical standard deviation of niche overlap, should also have a standardized normal distribution assuming all previous calculations and approximations exact. Figure 19 is a histogram of the 102 values thus obtained. The ratio 0/0 is taken as 0. The sample mean of these 102 numbers is 0.2781 and the sample variance 2.1015. This sample variance is significantly larger than the theoretical value of 1, even at the 0.001 level. However, even using the theoretical variance of 1, the sample mean of 0.2781 is not significantly different from the theoretical mean of 0.

Thus both the standardized sample mean and the standardized sample variance of niche overlap in the Monte Carlo simulations are, or are close to being, overdispersed in comparison with their respective theoretical variance of 1. In neither case is there evidence of bias in comparison with the respective theoretical mean of 0.

Several morals may be drawn from this cautionary tale.

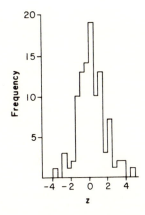

FIGURE 19. A frequency histogram of the transform z of a statistic that compares the sample mean of the number of niche overlaps in 100 pseudo-random food webs with the theoretical mean. There are 102 transformed values. If the pseudo-random number generator were random, the theoretical expectations were correct, and the normal approximations exact, this histogram would approximate a normal distribution with mean 0 and variance 1. (Based in part on Table 8.)

First, it is not safe to assume that the pseudo-random number generator or the seed for it supplied by the software of a computer system is unbiased in a particular application. Second, when simulation with a pseudo-random number generator is necessary, it is useful to have analytical results for checking the outcome of at least some of the simulations. Third, when the comparison of analytical and simulation results shows systematic bias, the damage that is done using the pseudo-random number generator depends on the direction of the bias. Fourth, there may exist linear combinations of the pseudo-random variates that are better behaved in some respects than the individual pseudo-random variates; e.g., in this application, the means of the standardized mean and standardized variance of niche overlap were not significantly different from their theoretical values of 0.

91

5.4 SUMMARY

The question, "Should we be surprised by the high observed frequency of interval food webs?" is made precise by specifying model universes from which to draw expectations. Six of these model universes refer to the food web matrix. One refers to the niche overlap graph. Artificial food web matrices or niche overlap graphs pseudo-randomly sampled from these model universes provide estimates of the expected frequencies of interval food webs or interval niche overlap graphs.

Except for the sink food webs, all sets of food web versions examined are interval significantly more frequently than expected from any of the models. The sink food webs are interval significantly more frequently than expected from the model assuming that niche overlap graphs are random graphs, but not significantly more frequently than expected from the six models of the food web matrix. The probable reason is that the parameters of the sink food webs specify regions in the six model universes where the frequencies of interval food webs are very high.

Since all of the sink food webs are consistent with a one-dimensional niche space in single habitats, the failure of the observed frequency of interval sink food webs to be significantly larger than expected from some models in no way weakens the conclusion that all or nearly all single-habitat community or sink food webs are interval.

A calibration of the Monte Carlo sample mean and sample variance of niche overlap against theoretically calculated values shows rough consistency. It is concluded that the high observed frequency of arrangements of niche overlap that can be represented in a one-dimensional niche space does not result from the operation, within the framework of several plausible models, of chance alone.

CHAPTER SIX

If Niche Space Is
One-Dimensional

Thought depends absolutely on the stomach, but in spite
of that, those who have the best stomachs are not the best
thinkers.

Voltaire, Letter to d'Alembert,
20 August 1770

6.1 WHAT IS THE ONE DIMENSION?

6.1.1 *Non-Uniqueness*

If a one-dimensional niche space suffices in general to represent the overlaps, along the trophic dimensions, of niches in a
single habitat, what is that one dimension?

Any answer to the question is probably not unique, in at
least two respects.

First, the single dimension, if it exists, may not be the same
from one habitat to another.

Second, within a given habitat, "similarity of species along
one dimension [often] implies similarity along another. This is
because the dimensions that ecologists recognize are rarely
independent" (Schoener, 1974, p. 32). For example, Schoener
(1967, p. 476) demonstrates a nondecreasing, and in a certain
range linearly increasing, relation between the length of
predatory individual lizards and the length of their prey
(Figure 20). If the niche overlaps of several species of lizards
could be represented along the single dimension of prey length,
and if the relation between prey length and predator length
were as shown in Figure 20, then the niche overlaps of the
several species could also be represented along the single dimension of predator length.

93

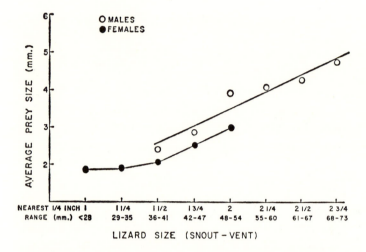

FIGURE 20. Average prey size as a function of predator size for male and female lizards. (Schoener, 1967, p. 467, by permission of the author and publisher, © 1967 by the American Association for the Advancement of Science.)

(If the niche overlaps could be represented along the dimension of predator length and if the relation between predator length and prey length were as shown in Figure 20, it is not necessarily true that the niche overlaps could also be represented along the dimension of prey length. Two species of lizards that might be represented by nonoverlapping intervals of predator length in the region where the curve in Figure 20 is horizontal would necessarily overlap in average prey size.)

In general, if one measurement can serve as the dimension representing niche overlap in a niche space, then any other measurement that is a strictly increasing or strictly decreasing function of the first measurement can also serve as that dimension. Therefore if $x_1, \ldots, x_d$ are a set of measurements that are positive (possibly after change of sign or rescaling), each of which could serve to represent niche overlap, then any linear combination $\Sigma a_j x_j$ with strictly positive coefficients a_j could also serve as the dimension of niche space. Consequently, there

94

is no formal way to distinguish one dimension of a niche space from a manifold of strictly monotonic functions of that dimension.

6.1.2 *Suggestions from the Food Web Studies*

Some studies of food webs identify candidates for the single dimension of niche overlap. Because it has not been common practice among field ecologists to coordinate the study of food webs with the study of niche boundaries or resource partitioning, only a few of the food web studies have appropriate data.

Hairston's (1949) four species of *Desmognathus* salamanders have all six possible pairwise overlaps in diet (food web 4), even if the food items are restricted to those coming from a forest habitat. The species appear to segregate along the dimension from aquatic to terrestrial. This dimension is often measurable as the distance from the nearest stream. The diets of the four species change broadly with increasing terrestriality. The present approach would accept terrestriality as one possible candidate for the dimension of trophic niche space if, for every pair of species, there were at least one point on the dimension of terrestriality where the two species in the pair overlapped. Another candidate, perhaps less plausible on the available evidence, is altitude. Figure 21 shows the ranges of altitudes at which each species of salamander was found. Within the Black Mountains Area viewed as a whole, there are all six possible pairwise overlaps in altitudes. Within individual coves, the species separate by altitude. But the 10 specimens of each species that were studied to form the food web matrix were not similarly distinguished according to cove of origin. A strong test whether altitude determines niche overlap would be to see whether dietary overlaps within each cove faithfully reflect altitudinal overlaps.

The interval food web 8.1 of species of *Conus* on marine benches (Kohn, 1959) reports no overlap in diet between *C. chaldaeus* and *C. flavidus*. Both species are found between 50

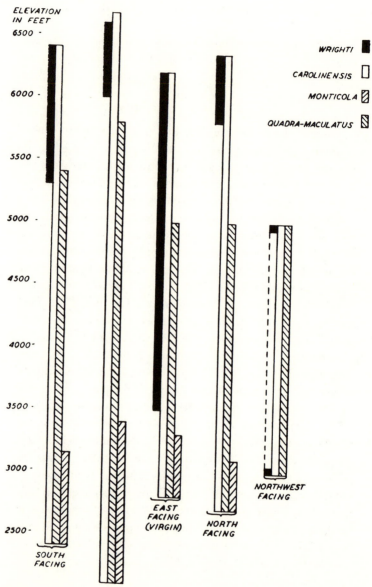

ELEVATION
IN FEET

WRIGHTI ■
CAROLINENSIS □
MONTICOLA ▨
QUADRA-MACULATUS ▧

SOUTH FACING

EAST FACING (VIRGIN)

NORTH FACING

NORTHWEST FACING

FIGURE 21. The vertical distribution of species of *Desmognathus* in the Black Mountains Area, based on observations from individual coves. (Adapted from Hairston, 1949, p. 56, by permission of the author and publisher.)

96

and 60 percent of the distance from the shore to the seaward edge of the marine bench at Station K1 (Kohn, 1959, p. 62). Therefore that distance measure can be ruled out as the one dimension of niche space. On the other hand, *C. sponsalis*, *C. abbreviatus*, *C. ebraeus*, and *C. chaldaeus* have all possible pairwise overlaps of diet on the marine benches. In all four species, individuals between 27 and 28 mm long were found on the marine bench at Station 5 (p. 55). Assuming that the dietary overlaps found from the pooled marine bench samples are faithfully reflected at Station 5, the length of the snails is then a candidate for the single dimension representing niche overlaps.

The adequately sampled species of *Conus* on reef platforms (food web 8.3) are interval, as described in section 3.2. The diets of *C. ebraeus* and *C. sponsalis* overlap, but neither diet overlaps with that of *C. flavidus* or *C. lividus*, which do overlap with each other. Since all four species are found between 0 and 30 percent of the distance from the shore to the outer edge of the reef platforms at stations 3, 7, and 9 (Kohn, p. 63), that distance measure can again be ruled out as the one dimension of niche space.

Paine's (1963) interval food web of eight predatory sympatric gastropods (food web 12) cannot have shell length as the single dimension of niche space. The diet of *F. tulipa* overlaps the diets of the seven other gastropods, but the shell lengths of *F. tulipa* are strictly smaller than those of *Pleuroploca* and strictly larger than those of *Murex* (p. 66). If Paine had reported only the average shell lengths of each species and their standard deviations, it would not have been possible to draw this conclusion about overlap of shell lengths.

6.1.3 *Suggestions from Other Studies*

In a colossal analysis of 81 studies of niche relations in groups of three or more species, Schoener (1974) tabulates and ranks by importance the dimensions along which species differ in their uses of resources. Observing that "dimensions within the

97

broad categories of food, space, and time tend to be the most correlated," he uses, in one of two analyses, "only one dimension from each of the three categories" to find the frequency distribution of number of dimensions per study. "For a maximum of three dimensions, two is by far the commonest value. Even if one admits the imprecise identification of important dimensions, separation appears generally to be multidimensional" (p. 29).

The food available may well depend on the hour or season, and spatial migrations often depend on the hour or season. So even the limitation to one dimension from each category of food, time, and space does not guarantee that the resulting dimensions are functionally independent, and Schoener makes no such claim.

If the dimensions are functionally independent, Schoener's finding (1974) of more than one dimension can still be reconciled with the one-dimensionality of niche space in single-habitat communities if he is looking at composite communities (section 2.3). It is difficult to know in general whether he would distinguish as different microhabitats or as different macrohabitats what we would call different habitats, because neither his concepts nor ours are unambiguously defined. At least one example suggests some correspondence between his concepts and ours, however. Schoener describes Kohn's (1959) study of *Conus* in Hawaii as reporting a macrohabitat differentiation according to marine formation and a microhabitat differentiation according to substrate. We have seen (section 3.2, food webs 8.2, 8.3) that distinguishing between reef and marine bench habitats does affect the dimensionality required to represent niche overlap.

Cody (1968, p. 134) provides unintentional support for the one-dimensionality of niche space within a single community. In comparing niche separation in 10 grassland bird communities, he explains that he averaged pairwise comparisons of species within each community "because species are often linearly arranged along an environmental variable. That is, if species A, B, and C are distributed in sequence along, for instance, a bill-length coordinate, the difference between pair A-C is necessarily greater than that between pairs A-B and B-C, which thus introduces scatter into the resulting values of pair-wise differences. A community mean by reducing this type of variation conveys an averaged and simplified idea of interspecific differences."

Gilpin (1977) has performed competition experiments in the laboratory among all possible pairs from a collection of 30 species of *Drosophila*. A species' success in each experiment is assessed by the proportion of all surviving individuals at the end of the experiment that belong to that species. The outcomes of these experiments can be very well described by assuming that all species are ranked on some single dimension and that one species will completely displace another if its rank on this dimension exceeds the rank of the other. This study incorporates all the factors that result in one population replacing another in the laboratory. In Gilpin's view, among those factors some, such as resource utilization and competition, may correspond to components of the one dimension inferred here from food webs; other factors incorporated in his one dimension, perhaps allelopathic and behavioral, may be omitted from ours.

Candidates abound for a single dimension to describe niche overlap, though they have not unfortunately always been studied in relation to feeding overlap. A possibility that may be general but is difficult to measure is that the single dimension is a benefit-to-cost ratio such as the energy gained by a predator per unit energy expended in predation. This possibility could relate these food web studies to studies of energetics. Among other possibilities, Terborgh reports the distribution of foraging height of four species of antbirds *Myrmotherula* (May, 1973, p. 168; Pianka, 1976, p. 134; MacArthur, 1972, p. 43). Mac-Arthur (pp. 75–76) graphs Storer's data on the distribution by weight of food of three species of North American *Accipiter*. The distribution by depth, temperature, and dissolved oxygen of three species of game fish in three TVA impoundments appears in Figure 22.

What is needed is the coordination of the empirical study of dietary overlaps and differences with the study of measurable dimensions of predators, prey, and environment. It is desirable to recognize that the dimension or dimensions ultimately defined by such a study have a different operational definition from the dimension or dimensions defined operationally by competition experiments such as Gilpin's or by resource partitioning studies (Cody, 1968; Schoener, 1974). It is therefore not to be expected that the results of these different kinds of studies will always coincide, nor that the term "dimension" will be clearly understood without an explicit operational definition.

6.2 WHAT FOLLOWS FROM ONE DIMENSION?

A one-dimensional niche space has been assumed in mathematical models of the distribution of species' abundances (MacArthur, 1957, 1960) and of niche overlap in relation to resource use (May, 1973, 1974).

These models assume a single dimension as a matter of analytical convenience, rather than as a strong conclusion of empirical study. For example, after an extensive discussion of a one-dimensional niche space, May (1974, p. 318) introduces the multidimensional problem thus: "Although there are circumstances in nature where a one-dimensional model is at least a good approximation (see, e.g., the results of Cody (1968) for grassland bird communities), more usually there will be several relevant resource dimensions."

These models also supplement the assumption of a one-dimensional niche space with other assumptions. The details of these other assumptions strongly influence the predictions

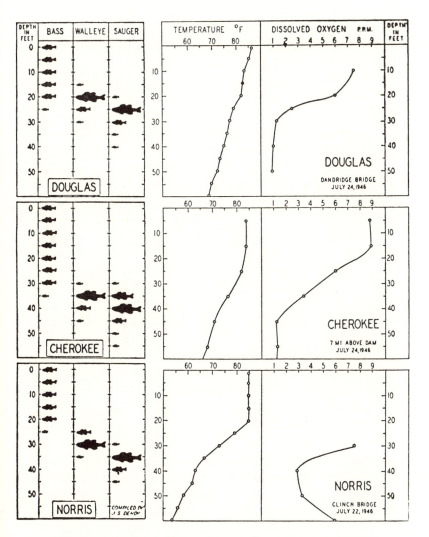

FIGURE 22. The distribution by depth, temperature, and dissolved oxygen of three species of game fish in three TVA impoundments in midsummer. (Based on Dendy, from Odum, 1971, p. 155, by permission of the author and publisher.)

101

of the models. Consequently, the conclusions of these models follow from one-dimensionality in conjunction with other assumptions, but not from one-dimensionality per se. The arguments and evidence in the preceding chapters for a one-dimensional (trophic) niche space support one of the several assumptions in each of these models, making it possible to attend more closely to testing the remaining assumptions.

We now review some details of two models of MacArthur (1957, 1960).

MacArthur's (1957, 1960) model of non-overlapping niches predicts correctly the distribution of species' abundances under special circumstances. It is qualitatively contradicted by the overlapping of niches observed in every real food web. The same distribution of species' abundances is predicted by a balls-and-boxes model (Cohen, 1966). This model allows niche overlap to occur as a result of the random occupancy of subniches by individuals of different species. H. Ronald Pulliam (personal communication) points out that the balls-and-boxes model could easily give rise to non-interval niche overlap graphs. If every niche overlap graph in a single habitat is interval, then the balls-and-boxes model cannot apply. Whether the balls-and-boxes model is compatible with the observed high frequency of interval niche overlap graphs could be determined by Monte Carlo simulations.

MacArthur's model of overlapping niches fails to describe adequately the distribution of species' abundances, even where his model of non-overlapping niches succeeds. A further empirical test of the model of overlapping niches is possible with the present food webs. The expected number of overlaps among the niches of n predators is $(2/3)\binom{n}{2}$ or $n(n-1)/3$, according to Appendix 3, which states the model formally and carries out the calculations. Hence, using the data in Table 5, a plot (Figure 23) of the number E of edges in the niche overlap graph against $\binom{n}{2}$ should yield a straight line through the origin with slope $2/3$.

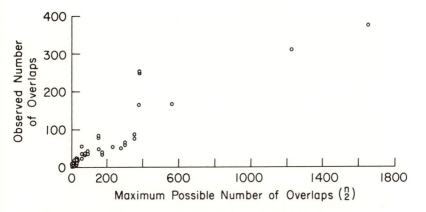

FIGURE 23. The observed number of niche overlaps, or edges in the niche overlap matrix, as a function of $\binom{n}{2}$, where n is the number of kinds of predators, for all food web versions. The observed slope is lower than the slope of 2/3 predicted by MacArthur's model (1957, 1960) of overlapping niches. (Based on Table 5.)

The least squares estimate of the slope, using all food web versions, plus or minus its standard deviation, is 0.25 ± 0.02. For the version A food webs, the slope and standard deviation are 0.27 ± 0.02, and for version B food webs, 0.24 ± 0.02. In all three cases, the linear correlation coefficient exceeds 0.9. Thus the relation between E and $\binom{n}{2}$ is close to linear, as the model would predict, but the observed slope is considerably lower than the predicted.

The discrepancy between predicted and observed overlaps could be made up by overlaps along unobserved nontrophic dimensions, since food webs give only overlaps along the trophic dimension(s). We have no evidence to evaluate this possibility.

As might be expected from previous differences between community and sink food webs, the two classes also differ in the relation between E and $\binom{n}{2}$. The slope of the regression and its standard deviation for community food webs are

0.24 $\pm$ 0.03, with a linear correlation coefficient of 0.89. For sink food webs, the slope and standard deviation are 0.44 $\pm$ 0.03, with a linear correlation coefficient of 0.95. In both classes of food webs, the slope remains well below the predicted 2/3.

This analysis contributes one further bit of evidence against the empirical value of these two models of MacArthur (1957, 1960).

6.3 WHY ONE DIMENSION?

At least six interpretations are possible of why we have found the niche space of single habitats to have one dimension. The first three of these would reduce the result to a triviality. We reject them as insufficient. The last three are more interesting, but still far from satisfactory.

The least flattering possibility is that the finding is a tautology. According to this interpretation, in the absence of an explicit algorithm to decide when a community resides in a single habitat or in a composite of several habitats, we call communities composite when their niche spaces turn out not to be one-dimensional.

Our response is to admit that the definition (section 2.2) of what constitutes a single habitat is unfortunately less explicit than the procedure for determining when a food web is interval. Lindeman (1942) laments the multiple meanings that have been attributed to the term "habitat." Nevertheless, in every case where we interpret a food web as describing a composite community, the article reporting the food web provides objective justification. If we were able to fabricate multiple habitats at will, we would not have the embarrassment of the non-interval food webs 28.2 and 28.3, which very probably do pertain to single habitats and which we suppose (section 3.2) to be based on inadequate sampling of diets. Finally, even if one considers all food webs in our sample, or all version A or all version B food webs (Chapter 5), without regard to whether the habitats

104

are single or composite, there are far more interval food webs than could be explained by chance. Even if the accusation of tautology casts doubt (which we deny) on our hypothesis that all single-habitat food webs are one-dimensional, it cannot eliminate the excess frequency of interval food webs observed in comparison with expectations from several random models. So we dismiss this first interpretation.

Second, suppose that the underlying dimension is some natural continuous variable, such as prey size, seed hardness, distance from shore, altitude, salinity or humidity, or some positive linear combination of these. If a predator can take prey at two different values of this variable, it is plausible to expect it to be able to take prey at all intermediate values of the same variable. Hence on any given dimension, a predator's utilization should cover an interval, without holes. We accept this argument. But it leads only to the conclusion that a niche should be convex. It does not explain why *one* dimension is sufficient. It does follow if niches are convex (section 7.1.1) that three independent dimensions are always sufficient to represent niche overlap.

Third, any food web in which at least one kind of prey is eaten by all predators is interval, since the niche overlap graph is complete (it has all possible edges). (A food web with a kind of predator that eats all available kinds of prey need not be interval.) Eight of the 31 different food webs in our sample have at least one kind of prey eaten by all kinds of predators (food webs, 10, 16.1, 16.3, 19, 20, 23, 26, and 27). On the basis of this observation, suppose that there is an implicit or explicit bias to select as the vertices of food webs just those collections of organisms that have at least one kind of prey eaten by all kinds of predators. According to this interpretation, the un- usually high frequency of interval food webs is explained by this bias in selecting communities.

This interpretation cannot account for the unexpectedly high frequency of interval community food webs, since all eight of the above food webs are sink food webs and none of the community food webs has a prey taken by all the predators. Moreover, suppose we omit the eight food webs listed above and repeat the comparison of the observed number (6) of interval food webs with the number expected, among the eight remaining sink food webs (food webs 4, 5, 8.1, 8.2, 8.3, 12, 16.2, and 29). If S_A means the versions A of these eight food webs and S_B means the versions B, then from equations (5.1 to 5.3) we find

$$z_{S_A 6} = 1.72, z_{S_B 6} = 3.06, z_{S_A 7} = 3.92, \text{ and } z_{S_B 7} = 5.44.$$

The four corresponding z values using all 16 sink food webs, from Tables 10 and 12, are 1.74, 2.99, 3.92, and 5.44. Omitting the sink food webs containing at least one kind of prey taken by all kinds of predators makes little or no difference to our previous conclusions concerning sink food webs.

Though this interpretation is insufficient to account for our results, it does raise the interesting question, to which we offer no answer, why there are so many sink food webs with a prey taken by all predators. Three of the eight food webs with such prey have fewer than four predators. In those webs such prey could easily arise by chance alone. Seven of the eight food webs with such prey have six or fewer predators. We have not investigated whether chance alone suffices to account for the presence of such prey.

A fourth interpretation is that single-habitat communities that have one-dimensional niche spaces have stability that communities with higher dimensional niche spaces do not have. We shall show that there is no necessary connection between the one-dimensionality of a community's niche space and qualitative stability (May, 1973, pp. 70–74). This proof does not rule out the possibility, which we have not investigated, of a statistical association between qualitatively stable and interval food webs.

May (1973, p. 69) pictures a food web that is necessarily interval, since it contains only two predators, but that is not qualitatively stable.

Conversely, Figure 24 is a non-interval food web, since its niche overlap graph contains an asteroidal triple (Lekkerkerker and Boland, 1962). To prove that this could be the food web of a qualitatively stable community, let $w = (w_{ij})$ be the food web matrix of this food web graph: $w_{ij} = 1$ if there is an arrow from i to j, $w_{ij} = 0$ otherwise. From w we construct the matrix called A by May (1973, p. 71). If $y = (y_i)$ is the vector of deviations from hypothetical equilibrial values of abundances or biomasses of the 13 kinds of organisms in the food web, then A specifies the dynamics of community composition according to $dy/dt = Ay$. We construct A from w by formalizing May's (p. 72) practice:

(i) $a_{ii} = -1$. (Assume that each kind of organism is self-regulating.)

(ii) For $i \neq j$, if $w_{ij} = 1$ and $w_{ji} = 0$, then $a_{ji} = 1$ and $a_{ij} = -1$. (If organism j eats organism i, but not vice versa, assume that the favorable effect of i on j and the unfavorable effect of j on i dominate all other indirect effects.)

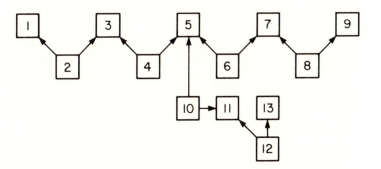

FIGURE 24. A hypothetical non-interval, qualitatively stable food web graph.

107

(iii) For $i \neq j$, if $w_{ij} = 0$ and $w_{ji} = 0$, then $a_{ji} = a_{ij} = 0$. (If there is no direct feeding relation between i and j, then i and j do not interact at all.)

The case $w_{ij} = w_{ji} = 1$, for $i \neq j$, which would be difficult to translate to a linear model, does not arise in this example. We leave its translation undefined.

The 13×13 matrix A constructed according to (i, ii, iii) satisfies the conditions quoted by May for qualitative stability. When some species may not be self-regulating, contrary to our assumption (i), the conditions quoted by May are not equivalent to qualitative stability. Conditions that are equivalent in the more general cases are known (Jeffries, Klee, and van den Driessche, in press).

The possibility that communities with interval food webs may have other desirable kinds of stability is not foreclosed. For example, a fifth possible interpretation of both our finding interval food webs and Schoener's (1974) finding two-dimensional niche spaces is suggested by a classic theorem about random walks (Spitzer, 1964, p. 83). A genuinely d-dimensional lattice-valued random walk is always transient when $d \geq 3$, whereas a random walk with 0 drift and finite absolute first moment is recurrent in one dimension and a random walk with 0 drift and finite absolute second moment is recurrent in two dimensions. The dimensionality of the space alone makes a qualitative difference in the possibility of recurrence.

One can create a relation between this mathematical curiosity and our empirical findings by viewing the time-trajectory in niche space of each kind of organism as a simple symmetric random walk. Define a community as a finite set of kinds of organisms whose simultaneous presence at a single lattice point, once observed, recurs with probability 1. Then it is possible for a community to exist in a one- or two-dimensional niche space, but impossible for such a community to exist in a niche space of dimension three or more.

108

Since we have no evidence that a simple symmetric random walk is an appropriate, or even plausible, model for the dynamics of niches in niche space, this interpretation establishes no more than a possibility.

The finding that single-habitat food webs are (usually) interval while trophic niches are commonly described in multidimensional terms may finally reflect the difference between community ecology and physiological ecology. Organisms may have more degrees of freedom in their physiological capacities to exist under varied circumstances than the biotic, especially trophic, interactions with other kinds of organisms in their community permit them to enjoy. This sixth interpretation might be put to empirical test by comparing the trophic dimensionality of natural or artificial communities which have varying intensities of biotic interaction, measured in some appropriate way. Without some more exact operational definitions, this interpretation is unfortunately vague.

In conclusion, we have no fully satisfactory answer to the question, Why one dimension? but we can clearly reject some of the interpretations that would reduce the result to a triviality.

6.4 SUMMARY

If a one-dimensional niche space can represent the overlaps, along the trophic dimensions, of niches in a single habitat, the single dimension identified in one community may differ from that in another. In a single habitat, the one dimension may be chosen from a manifold of monotonically related dimensions.

A few food web studies provide enough information on feeding and distribution to suggest what the one dimension may be. Among salamanders (Hairston, 1949), dietary overlaps may be represented by overlaps in distance from water or in altitude. Among snails (Kohn, 1959), dietary overlaps may be represented by overlaps in body length. Among the many other candidates for the one dimension of niche space, some can be rejected by data.

Different kinds of studies of niche space, such as those of resource partitioning or those based on competition experiments, employ different operational definitions of "dimension." A concordance among the results of the different kinds of studies would represent a major empirical discovery. If a concordance among the different operational definitions of "dimension" is taken for granted but turns out to be contrary to fact, the word will become a conceptual trap for the unwary, as have other equivocal ecological terms in the past.

The arguments and evidence for a one-dimensional trophic niche space support the assumption of a one-dimensional niche space in models of MacArthur (1957, 1960), May (1973, 1974), and others, making it possible to direct attention to testing the remaining assumptions of those models.

Several interpretations are possible of why we have found the niche space of single habitats to have one dimension. We argue against the adequacy of three trivializing interpretations. We disprove the conjecture that there might be some necessary relation between one-dimensionality and qualitative stability. The possibility of recurrence in one- and two-dimensional random walks, but the impossibility of recurrence in three- or higher-dimensional random walks, may account for our finding of one dimension in single habitats and Schoener's (1974) finding of two dimensions in many communities. Niche spaces may have one dimension because biotic, especially trophic, interactions with other kinds of organisms in their community have in some way collapsed the possible variation inherent in organisms' physiological capacities. We do not claim to offer a fully satisfactory explanation.

CHAPTER SEVEN

Extensions and Critique

> . . . not all naturalists want to do science.
> R. H. MacArthur (1972, p. 1)

> They have vilified me! They have crucified me. Yes, they have even criticized me!
> Richard Daley (1977)

7.1 EXTENSIONS OF THIS APPROACH

When niche space is not one-dimensional, the analysis of the combinatorial structure of food webs extends to representations of niche overlap in higher dimensional spaces and to approximations of a niche overlap graph by an interval graph.

7.1.1 *Higher Dimensions*

Let S be a collection of a finite number n of nonempty sets. For example, let each set in S be the set of prey organisms taken by one of n predators. The intersection graph G of S (Klee, 1969) has n vertices, one corresponding to each set in S. Two vertices of G are joined by an edge if and only if the two corresponding sets in S have a nonempty intersection. In our example, the intersection graph is just the niche overlap graph.

A food web is defined to be interval when its niche overlap graph is isomorphic to the intersection graph of a collection of intervals. To extend this approach to higher dimensions means to ask, when a food web is not interval, whether there is some class of subsets of a higher dimensional Euclidean space such that every real niche overlap graph is isomorphic to an intersection graph of some subsets in that class, and if so, to find the minimal necessary dimension of the Euclidean space. Independent evidence should be adduced that the class of subsets

111

of the higher dimensional Euclidean space corresponds to the measured form of real niches.

The only assumption we make in one dimension is that a niche is connected. Not every undirected graph is the intersection graph of curves in two-dimensional Euclidean space (Ehrlich, Even, and Tarjan, 1976). Every graph is the intersection graph of connected regions in Euclidean 3-space (Dewdney, 1977). Hence, if any connected region could be a niche, one or two dimensions may not be sufficient, but it is never necessary for niche space to be more than 3-dimensional in order to represent trophic niche overlaps.

In one dimension, the assumption that a niche is connected is equivalent to the assumption that it is convex. However, not every graph is isomorphic to the intersection graph of convex sets in two dimensions, according to results of Wegner reported by Klee (1969, p. 812). The class of graphs that are isomorphic with the intersection graph of some finite family of convex regions of the Euclidean plane includes all graphs with less than nine vertices and planar graphs, but not any nonplanar graph in which all neighbors of any vertex of valence (or degree) greater than 2 are of valence 2. Finally, according to Wegner, all graphs are isomorphic with the intersection graphs of convex regions in Euclidean 3-space. So any observed niche overlap graph can be represented by the overlaps of convex niches in Euclidean 3-space.

If an organism's range on one dimension is independent of its range on every other dimension, then the niche is a box with sides parallel to the coordinate axes. This is a special case of a convex niche. The minimum number of dimensions necessary to find an intersection graph of boxes that is isomorphic to a given graph is defined (Roberts, 1969a) as the "boxicity" of that graph. For a graph on n vertices, the maximum possible boxicity is the largest integer less than or equal to $n/2$ (p. 309). Thus the niche overlap graph of a food web with 10 predators can always be represented as the intersection graph of boxes in

5-dimensional space. Gabai (in press) develops upper and lower bounds for the boxicity of a graph.

Direct evidence regarding the form of niches is scattered. Krebs (1972, p. 173, using data of Birch) gives approximate contours of the Malthusian parameter r for two species of grain beetles as a function of the temperature and moisture content of the wheat in which the beetles live. These niches are convex but not rectangular. Krebs's later picture (p. 232) of two hypothetical niches as boxes in a plane with the same coordinates is perhaps only a pedagogical device. Additional evidence argues against the dimensions of niches being independent. Gates's (1969, p. 125) "climate diagram" for a cardinal based on actual values of metabolic and water loss rates shows a non-rectangular region in a plane whose coordinates are air temperature and radiation absorbed. Chapman (1931, pp. 106–21) and Allee et al. (1949, pp. 206–15) each devote a chapter to the

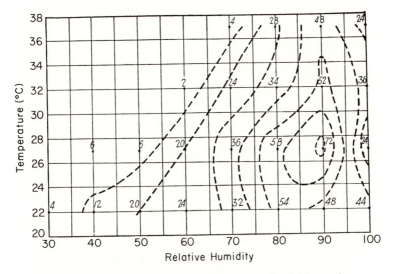

FIGURE 25. The percentage of grasshopper eggs that hatch (the numbers on the grid) as a function of the relative humidity and the temperature. (Based on Parker, from Chapman, 1931, p. 110.)

113

interaction of niche-limiting factors, principally temperature and moisture (e.g. Figure 25). In a space with the dimensions of salinity, temperature, and depth, shrimp are found in regions (Figure 26) that are not even convex. Based on the data of earlier authors, Maguire (1973, p. 238) points out, among other examples, that the fourth zoeal (larval) stage of the crab *Sesarma cinereum* has a minimum of mortality when moderate salinity is combined with either high temperature or low temperature, and an intervening ridge of higher mortality at moderate temperature and moderate salinity. With a cut-off above 20% mortality (not survival, as in the caption of Maguire's Figure 19), the niche of the fourth zoeal stage would not be connected in the salinity-temperature plane, though it might well be connected in a space with additional axes.

While it is worth developing further the theory of intersection graphs of various regions including boxes in d-dimensional Euclidean space, it is necessary to have some quantitative in-

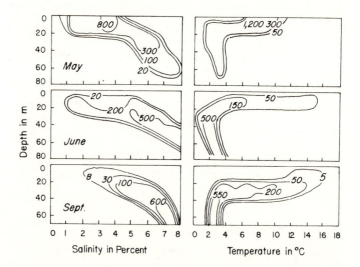

FIGURE 26. The abundance of the crayfish *Limnocalanus grimaldii* per m³ in the Gulf of Finland as a function of salinity, temperature, depth, and season. (Based on Bityukov, from Naumov, 1972, p. 98.)

114

formation about the shape of niches before applying this theory to particular cases.

If we assume that niches are convex regions, some remarkable inferences about niche space result from the application to the niche overlap graph of Helly's theorem and its many relatives. Helly's theorem states that for any finite collection of at least $d + 1$ convex sets in Euclidean d-space, if each family of $d + 1$ of these sets has a common point, there is a point common to all members of the collection. Danzer, Grünbaum, and Klee (1963, pp. 126–27) explicitly discuss the relation between interval graphs and Helly's theorem. Danzer and Grünbaum (1968) establish properties of boxes in Euclidean d-space that are sharper than those implied by Helly's theorem.

In the case of one-dimensional niche space, Helly's theorem applies to a niche overlap graph that is the complete graph on n vertices, since it assumes that the diets of every pair of predators overlap. The theorem says there is some point in niche space which is in the (trophic) niche of every predator. Indeed, it is obvious that if $[a_1, b_1], \ldots, [a_n, b_n]$ are the n intervals representing the niches of the n predators, then the interval $[\max_i a_i, \min_i b_i]$ is common to all the intervals. The theorem is less than obvious in higher dimensions.

Victor Klee (personal communication) points out that if niches are boxes (as usual, with sides parallel to the coordinate axes), then there is some point common to all the niches if and only if each pair of boxes intersect. Thus when all dimensions are independent, this statement, applied to multidimensional niches, is of the same form as Helly's theorem for one-dimensional connected niches.

7.1.2 *Nearly Interval Food Webs*

When a food web is not interval, one way to measure how far it is from being interval is by the minimal number of directed edges that must be added to the food web graph to render the food web interval. Another measure, which is not in general equivalent to the first, is by the minimal number of undirected edges that must be added to the niche overlap graph to make that an interval graph. Both measures are well-defined since a complete graph is an interval graph. We do not know of algorithms for calculating either measure short of exhaustive enumeration.

We would not consider it useful to count the number of edges that have to be removed from a non-interval graph to render it interval. Each such removal would amount to denying the existence of real data, whereas counting additional links could be interpreted as measuring the number of additional trophic relations whose existence has not yet been observed.

If a (niche overlap) graph is not interval, it may still be one-dimensional in the weaker sense that it is a subtree graph, that is, the intersection graph of a family of connected subsets of an undirected tree (a suggestion due to Victor Klee). A graph is a subtree graph if and only if it is a chordal (or rigid-circuit) graph, that is, every simple circuit with more than three vertices has an edge connecting two nonconsecutive vertices (Gavril, 1974). Figure 5 above is an example of a graph that is not a chordal (and therefore not a subtree) graph, though it would be chordal if there were an edge between vertices [1] and [3] or between [2] and [4]. Every interval graph is chordal, but not conversely. Efficient algorithms for testing whether a graph is chordal are known (Booth and Lueker, 1976, pp. 371–72).

Kendall (1969) proposes another approach that merits an ecological translation. We first describe his proposal formally, and then interpret it. Let w be an $m \times n$ food web matrix. Let A be the $c \times n$ dominant clique matrix (defined in section 2.1) formed from w; this is the matrix Kendall (1969) calls A and is not to be confused with the matrix A in section 6.3. Here c is the number of dominant cliques. Let $S = AA^T$, where A^T is the transpose of A. As a $c \times c$ symmetric matrix, S can be interpreted as a similarity matrix: s_{ij} measures the similarity between clique i and j. Apply Shepard-Kruskal multidimensional scaling in two dimensions to S.

If the food web w is interval, then A has the consecutive 1's property, that is, some permutation of the rows of A produces a $c \times n$ matrix B such that in every column of B there are not two 1's separated by a 0. Kendall (1969) calls any matrix of 0's and 1's that has this property of B a Petrie matrix. Kendall shows that if B is a Petrie matrix, then, starting from the main diagonal of $R = BB^T$, the elements of R never increase if one moves to the left or downwards. Kendall calls a matrix with this property of R a Robinson matrix. In a Robinson matrix R, the similarity r_{ij} between i and j is a non-increasing function of the difference between them. Since S is just a row- and column-permutation of a Robinson matrix if w is interval, the multidimensional scaling procedure in the plane should yield a set of points, each one corresponding to a clique, which fall in sequence along or near a smooth curve. (In practice [Kendall, 1970], this sequence lies along a horseshoe-shaped curve.) Each predator is then identified with the interval along this sequence which contains all the cliques to which the predator belongs. Those cliques will fall in one connected interval because B is a Petrie matrix.

When w is not interval, $S = AA^T$ still may be interpreted as a similarity matrix of the dominant cliques in A. If the multidimensional scaling procedure based on S produces a nearly one-dimensional arrangement of the c cliques, then we might say that w is "nearly" interval. An objectively measurable interpretation of the interval(s) associated with each predator is still needed.

A significant extension of these approaches to non-interval graphs would weight the distance that a food web is from being interval according to the quantity of the flow through the directed edges that cause the food web to be non-interval.

117

7.2 CRITIQUE

We identify four principal weaknesses in our approach and results.

7.2.1 *Ambiguous Primitives*

Our primitive descriptive terms are not uniformly and un-ambiguously defined. These terms include those required to define a food web, such as "kind of organism" and the relation "eats"; those required to set the boundaries of what is included in a food web, such as "habitat" and "community"; and those required to define niche space, such as "niche" and "dimension."

Before lapsing into despair over the ambiguities of these terms, we draw courage from the historical examples of terms, such as "force," "energy," "species," and "gene," which have emerged from misty origins into scientific use. Nash (1963) illuminates the logically untidy ways such evolution occurs.

We shall not undertake the Sisyphean labor of explicating or legislating verbal meanings for all our primitives. The preceding chapters demonstrate that ecologists have found implicit operational definitions that permit coherent, reproducible descriptions of nature.

In reporting food webs, a "kind of organism" is defined, not necessarily in terms of a taxonomic unit such as "species," but functionally in terms of the other kinds of organisms that it eats and that eat it.

The burden of ambiguity thus shifts to the relation "eats." A report that a kind of organism A eats a kind of organism B depends on biological contingencies. For example, suppose A eats B during the daytime, in summer, when a kind of organism C that A eats when C is available is absent or in low abundance. Suppose that under other circumstances A does not eat B. (This example is realistic: Naumov [1972, p. 549] describes the great differences between the summer and winter food webs in the arctic. Many primates are known to feed on fruits in season.) Probably most ecologists would report in a food web that A eats B even if they reported in the accompanying text that A never eats B at night; that is, they would integrate eating over a diurnal cycle. If the academic calendar limited their field seasons to the summer, ecologists might implicitly interpret "eats" to mean "eats during the summer." Naumov's is the only report we know of a food web which is explicitly season-dependent. As for the dependence of A's eating B on the abundance of C, there is simply no natural way to represent this (and many other kinds of dependence) in terms of a single, static food web.

The casting of field observations into a food web is a substantial simplification and abstraction of them. Whether our generalizations about niche space will remain true when food webs take account of season and other contingencies can only be determined by finer reporting and analysis of data.

We have already discussed a meaning of "habitat" (section 2.2) and shall not attempt to do likewise for "community."

Niches are no less an abstraction than food webs. The concept of a "niche" once referred to a region in a multidimensional hyperspace where the fitness (however defined) of a kind of organism exceeded a certain threshold. Currently the term usually refers to a region where a kind of organism is found (Schoener, 1974, p. 27). This change in meaning is not surprising since it is easier to see where an organism is than to define and measure its fitness there. Again we temporarily assume the statistical adequacy of field observations and dwell on the conceptual ambiguities of this seemingly simple definition.

The boundaries of where a kind of organism A is found may depend on time and other contingencies as much as A's diet and predators. There are additional problems in delimiting a niche. For example, it may be descriptively economical to say that A is normally distributed along the dimension of temperature, say, with some finite mean and variance. Then A's tem-

119

perature range is infinite in both directions. An alternative is to accept some basically arbitrary threshold below which the probability of occurrence of A will be replaced by 0. If one rejects the descriptive economy of a normal distribution, A's range of temperature may increase with sample size. Thus with or without the normal distribution as a descriptive convenience, the data may lack the neatness of the fixed intervals by which we represent niches here. A theory based on utilization functions (e.g. May, 1975) that is insensitive to differences between zero and small positive numbers avoids many of the difficulties inherent in the concept of a discrete niche.

7.2.2 *Statistical Shortcomings of the Data*

We pointed out in section 2.2 that an undirected graph can be changed from interval to non-interval, and from non-interval to interval, by the addition of a single edge. This sensitivity of interval graphs means that it is important to be able to make quantitative statements about the probable completeness of a reported food web graph.

As a first step in this direction, it would be highly desirable for a published food web to be accompanied by a "yield/effort" graph. The abscissa of this graph would measure the observer's cumulative effort at observation, whether by duration of observation during defined times of day or seasons of year, quadrats sampled, traps set, or stomach contents analyzed. The ordinate of this graph would measure the cumulative number of directed edges appearing in the food web graph, for each cumulated effort of observation. The graph would be a step function since the number of edges is an integer, and non-decreasing, but one could fair a smoothly increasing curve through this step function. The slope of the curve should decrease as the cumulative effort of observation increases and fewer relations remain to be observed.

If observations stopped while this yield/effort curve was still rapidly increasing, the food web derived from these observations would probably not be nearly complete. If observations stopped when the slope of the yield/effort curve was 0 or close to it, most eating relations would probably have been observed. Though many yield/effort graphs could probably be derived from the field notebooks which produced the food webs in our sample, we have never seen such a curve published. Robert T. Paine has students analyze their data in this way in field exercises, and notes the formal similarity of the idea to a species-area curve. Altmann and Altmann (1970, p. 116) implement exactly this idea in measuring the home range of baboon troops.

If one assumes that the number of kinds of organisms in the community is somehow known or bounded above, the number of possible edges in the food web graph is finite. As the study proceeds, sampling can concentrate on those possible feeding relations which have not yet been established. At the end of observation, it would then be possible to establish an upper confidence limit (the lower limit being 0) on the frequency of feeding relations which were not observed.

When the number of kinds of organisms in the community must be determined from the same sample being used to estimate the food web, the number of directed edges in the food web graph and the frequency of feeding associated with each of them are confounded. It is then possible to estimate an upper bound for the total frequency of feeding along all undetected directed edges, without bounding the number of such edges. This situation also arises in insect trapping or vocabulary estimation, where the number of species or words is unknown and the observer must attempt to estimate simultaneously the number of unobserved categories and their frequency.

Following the inspiration of the jackknife (Mosteller and Tukey, 1968; Miller, 1974), a technique for estimating the variation due to sampling in a given set of observations would be to divide the observations into some number k of roughly equal blocks and to examine the k food webs or niche overlap graphs obtained by leaving out of the sample in turn each one of the k blocks.

121

7.2.3 *Lack of Quantitative Content*

When food webs are reported as directed graphs or as matrices of 0's and 1's, they weight prey-predator arrows describing rare events equally with arrows describing frequent events. The present combinatorial approach will not answer quantitative questions about the flow in food webs of energy, nutrients, or toxic materials. Recent efforts devoted to these quantitative questions are massive (Levin, 1975, where there is a bibliography of bibliographies on pp. 292–93; Patten, 1975; Roberts, 1976a; Russell, 1975; Wiegert, 1976). Watt (1975, pp. 142–43) remarks, "we have now had about 40 years of fairly intensive activity devoted to building mathematical models of ecological phenomena such as predation, parasitism, herbivory, or abiotic effects. The vast majority of all the models put into the literature, my own included, either do not describe the phenomena they purport to describe, or contain internal mathematical problems . . . , or worse yet, have both difficulties."

Rather than review these efforts here, we are content to observe the potential importance of recent approaches to modeling that are intermediate between the purely combinatorial and the strictly quantitative. Roberts (1976b) calls such approaches "structural analysis" and discusses particularly his signed digraph technique, which is still in its mathematical childhood. It may shed light on the qualitative behavior of complex systems such as food webs.

In correspondence, Thomas W. Schoener and Robert M. May independently raise related objections to which it is impossible to respond with the available, largely dichotomous, food webs. In some situations, such as that of *Conus* in Hawaii, Schoener suggests, if enough predator specimens were taken, nearly all the predators would be found to prey on nearly all available prey. There would then be so many niche overlaps that the likelihood of an interval food web by chance alone would be high. He wonders if it is fair to include an arrow between a prey and predator if neither is quantitatively important to the other, although predation may occasionally occur. May asks whether, in some sense, the high frequency of interval food webs is an artifact that results from replacing underlying, continuous variables describing predation by 0's and 1's.

One way to answer these questions is to observe the food web of a community in quantitative detail, using explicitly planned sampling, and reporting the percentage each prey constitutes of each predator's diet and the distribution among predators of predation on each prey. Using various thresholds (such as: at least 1 percent of a predator's diet, at least 5 percent of the predation on a given prey, and so on), singly and in combination, one could construct food web matrices of 0's and 1's like the ones analyzed here, and examine the dimensionality of niche space inferred. It would then be possible to determine whether the results are invariant regardless of the thresholds used. If this procedure were performed for a variety of communities, it would also be possible to determine whether there are some thresholds that give more consistent results than others.

Highly laborious though such an exercise would be, it must be useful. If it shows that the unexpectedly high frequency of interval food webs depends on the choice of an appropriate combination of thresholds, but disappears otherwise, then the fact that ecologists have chosen such thresholds in the past demands explanation. Theories that have posited a one-dimensional niche space will have to be generalized to multidimensional theories. If the exercise shows an unexpectedly high frequency of interval food webs for a wide choice of thresholds, then our present conclusion that trophic niche space is one-dimensional is made much stronger. In either event, the work will yield much more systematic and detailed descriptions of the trophic dynamics of communities.

7.2.4 *Lack of Dynamic Theory*

Our argument that a single-habitat food web is specially structured for compatibility with a one-dimensional niche space has proceeded by enumeration of observed individual cases. To a mind of theoretical cast, this approach may be less persuasive than, or at least need to be accompanied by, a derivation of the conclusion from a more fundamental underlying theory. It would be preferable if such a theory were dynamic (time-dependent). For example, it would be nice to deduce that single-habitat communities must evolve toward compatibility with a one-dimensional niche space from the laws of Mendelian genetics and some especially plausible formalization of Darwinian evolution. Section 6.3 does not pretend to offer such a derivation.

There are two reasons for this shortcoming, aside from the need to establish general claims on the basis of observation rather than theory. First is the paucity, as Watt (1975, pp. 142–43) notes, of relevant credible theory to provide a starting point for deduction. Second is the static, rather than dynamic, nature of the data on food webs and niche overlap. Reports of food webs at different times for the same community, such as Naumov's (1972, p. 549) for the arctic, are exceptional. Reports of how food webs change continuously in time or geographically along gradients are unknown to us. In the absence of data about variation in time and space, any presumed equilibrium can be explained by multiple models. The elaboration of dynamic theories might best accompany, rather than precede, empirical study of changing food webs.

"If the dynamical laws governing the transformation and evolution of communities are to be discovered, it is essential to work with non-equilibrium systems through time either by a search for evolving communities or by a deliberate perturbation analysis of communities at present at stable points" (Lewontin, 1969, p. 23).

124

7.3 SUMMARY

In Chapter 6, we viewed the claims of the preceding chapters with the eye of faith, attempting to interpret, apply, and explain them. In this chapter, we view the same claims with less sympathetic conviction.

When niche overlaps are not consistent with a one-dimensional niche space, our combinatorial approach can reveal whether the niche overlaps could be represented by the overlaps of regions in a higher dimensional space. If niches are convex or connected regions, more than three dimensions are never required to represent niche overlaps. If a niche is a box in niche space with sides parallel to the coordinate axes, the dimension of niche space required to represent any observed niche overlap graph never exceeds half the number of kinds of predators. It is necessary to have quantitative information about the actual shape of niches before applying this theory to particular cases.

When a food web is not interval, it may also be worth examining how far it is from being interval. A technique for measuring the similarity between dominant cliques of predators can approximate an interval representation of the predators.

Our approach and results suffer from four major shortcomings. First, the very concepts in terms of which the data are reported and the results framed are ambiguous abstractions. Second, statistical features of the data used, especially the sampling design and reporting, leave much to be desired. Third, even if the concepts were clear and the statistics of the data impeccable, the claimed results do not attempt to answer many of the quantitative questions which are important to ecologists. Finally, a derivation of the claimed results from a more fundamental dynamic theory is lacking.

CHAPTER EIGHT

Review of Open Problems

The eyes of all wait upon thee,
And thou givest them their food
in due season.
Psalm 145:15

About 50 percent of the children
in the world do not receive
adequate protein nutrition.
A. M. Altschul (1967, p. 221)

8.1 INTRODUCTION

The summaries of Chapters 2 to 7 synopsize the approach and results of this work. This chapter collects some possibilities for future theoretical and empirical work. These open problems arise within the framework of the approach taken here. This approach, motivated by the desire to learn more than was being learned from the wealth of reported food webs, is limited. Whatever may be its successes, this approach cannot escape the discrete or combinatorial nature of its basic concepts. Some of the serious consequences of this limitation have been pointed out (section 7.2).

On the other hand, the present approach may usefully shake loose the attention of ecologists from the hypnotic gaze of some currently important theoretical concepts. These concepts have their own limitations as tools for gleaning knowledge from data. As the suggestions in section 8.3 will show, the present approach may also lead to the collection of data that go beyond the limitations of a purely combinatorial approach.

8.2 THEORETICAL

We shall distinguish between open theoretical problems that are scientific and those that are mathematical. The scientific

126

problems are unexplained phenomena found in data. They are subject to revision by future data. Their solutions may well include mathematical arguments. The mathematical problems are unsettled questions in pure analysis. They can be studied independently of data.

One scientific problem is to explain the apparent invariance in the ratio, near 3/4, of prey to predators in community food webs (section 4.2.2). If ecologists are more likely to lump prey species into one "kind" than they are to lump predator species, then the finding in a single food web that there are fewer kinds of prey than kinds of predators might not be a surprise. But the invariance in the ratio of prey to predators over many food webs would remain unexplained. Moreover, in many of the food webs studied here, many of the kinds of organisms are both predators and prey, and therefore subject to identical lumping. A behavioral model of predators, perhaps similar to Schoener's (1971), to account for the regularity in the "span of control," or number of kinds of prey taken per kind of predator, will probably have to be combined with an ecological model, perhaps similar to May's (1973, 1974), relating niche overlap to limited resources and competition.

Another scientific problem is to explain quantitatively the observed distribution of the length of maximal food chains (section 4.2.3). An explanation should be able to describe accurately the distributions of maximal food chain length in other real food webs and should have parameters with independently measurable and testable interpretations.

A third scientific problem is to account for the preponderance or universality of one-dimensional niche spaces for single-habitat communities. The attempts in section 6.3 do not claim to be satisfactory.

The mathematical problem of counting interval (undirected) graphs and interval food webs includes several possible approaches. János Komlós (personal communication) of the Mathematical Institute of the Hungarian Academy of Sciences has calculated the asymptotic probability that a random graph is an interval graph, using the methods of Erdös and Rényi (1960). Comparison with our exact and Monte Carlo results suggests that $n = 40$ vertices is not a large enough number for the asymptotic theory to be useful. Since the largest real food web (version 7.12) in our sample has 58 kinds of predators, there remains a range of n, from 7 to perhaps 10^2 or 10^3, for which results, exact or approximate, are still desirable. An approach more general than the very detailed counting arguments that lead to equations (5.4) is especially needed. The problem of enumerating directed graphs (food webs) that are interval has not yet been attacked analytically.

The conjectured inequalities of the means and variances of overlap in the food web models (section 4.3.1) remain to be settled.

Monte Carlo simulations of the balls-and-boxes model (section 6.2) could show whether the model is statistically compatible with the observed frequency of interval niche overlap graphs.

Even though there is no necessary connection between a food web being interval and a food web being qualitatively stable (section 6.3), it is worth studying whether, in some universe of food webs, those that are interval are more likely to be qualitatively stable than those that are not interval.

An algorithm for constructing all the interval graphs that have the least number of edges in addition to those of a given non-interval graph would be useful for measuring how far a given non-interval graph is from being interval (section 7.1.2). The same applies to non-interval food webs.

Though less well defined than the preceding problems, the area between the present purely combinatorial approach and traditional quantitative approaches deserves mathematical thought. Qualitative stability (section 6.3) and structural analysis (section 7.2.3) are examples of work in this area. More sophisticated tools could be used.

8.3 EMPIRICAL

The principal empirical problem arising from this work is to determine whether the one-dimensionality of the niche space of a single-habitat community is really universal when the community is adequately sampled, as we claim, or whether the apparent exceptions such as food webs 28.2 and 28.3 are genuine. If the latter is confirmed, we would like to know what distinguishes the communities with one-dimensional niche spaces from the others.

Rather than exhuming more published food webs that were not prepared with this investigation in mind, it would be preferable to exploit new opportunities for observation, reporting, and analysis. Food webs should be reported as several quantitative matrices, each showing, if possible, the actual number of specimens of each kind of prey taken by each kind of predator, under several specified conditions. The investigator may in addition derive percentages of row or column totals from these basic data, but the underlying quantitative details should not be thrown away in reporting. Except for the very simplest food webs, graphs alone are inadequate to convey the required information. They should be used only to give an overall impression, if at all.

The distinction between studies of food webs and studies of resource partitioning should be obliterated. The multidimensional range of conditions under which each predator feeds should be studied concurrently with the kinds of organisms it eats. Particular attention should be paid to the regions of overlap among predators. The distribution of predators along whatever niche dimensions are chosen should be reported in a way that enables readers to learn the actual numbers of individuals observed in each region of the niche. Ideally, such a report would include (possibly as an appended table) the full joint frequency distribution of all the predators and prey along

129

each of the observed dimensions of niche space. If the number of observations is large and the joint frequency distribution can be very well described by some frequency distribution with a small number of parameters (for example, if each kind of organism has a multivariate lognormal distribution in niche space and the different kinds of organisms are all independent), then at least the parameters, sample sizes, and actual sample boundary of the niche of each kind of organism should be reported. Merely reporting summary statistics such as mean and variance of where each kind of organism appeared on each dimension, one at a time, is useful only as a preliminary to thorough work.

By observing and reporting several niche dimensions jointly, one gains some information about the shape of the niche in a niche space of dimension greater than one (section 7.1.1), if a food web is non-interval.

Changes in feeding relations should be recorded over time, on the scale of days, seasons, or years. Observations indexed by time of observation make it possible simultaneously to tabulate yield/effort curves and display the time-dependence of food webs.

Yield/effort curves (section 7.2 2) should be reported wherever they are appropriate. "Yield" could be interpreted as the number of directed edges in a food web graph, and also as the number of niche overlaps (along various dimensions) in the concurrent study of niche space.

The variation in a given set of observations of diet or resource use could be assessed by means of the jackknife (section 7.2.2).

Simultaneously with time-indexed studies of feeding relations and niche occupancy or resource partitioning, perturbation experiments would be highly desirable (section 7.2.4). For example, if a baseline period of observation established that a food web was interval, the community might be perturbed so that the food web became non-interval. This might be done by interrupting an eating relation while leaving predator and prey present in the community (using an odorous chemical to render the prey unattractive, for example), or by removing either a predator or a prey organism. It could also be done by adding a predator or prey organism and altering eating relations or niche overlaps. The consequences could then be observed over time. It would be particularly interesting if the community, under its own dynamics, then reestablished its compatibility with a one-dimensional niche space. (After the unintentional introduction of the Amazonian piscivorous cichlid *Cichla ocellaris* into Gatun Lake, Panama, the "generalized" food web changed form radically [Zaret and Paine, 1973, p. 452] but remained, as it was before the introduction, an interval food web.) The converse experiment of changing a non-interval food web into an interval food web would be a natural companion. In both kinds of experiments, the minimum number of predators necessary to have a non-interval food web is four. Though these experiments might be expected to be complex, illuminating laboratory experiments with four different species have been carried out (Vandermeer, 1969). Experiments such as those proposed here are plausible for both the field and the laboratory.

Sample of Food Webs

The number at the top of each column indicates the predator, parasite, or other consuming kind of organism. The number at the left of each row indicates the prey, host, or other kind of organism consumed. The key with each food web matrix identifies the number. The entries of 0, 1, −1, and −2 in the food web matrices are explained for each case in Chapter 3.

| 1.1 | Bird, 1930 | 383 | Prairie, Canada | Community |

FOOD WEB MATRIX

	1	2	3	4	5	6	7	9	10	11	12	13	14	15
1	0	1	1	0	0	0	0	0	0	0	0	0	0	0
4	0	1	0	0	1	0	1	0	0	0	0	0	0	0
5	0	1	1	0	0	0	0	0	0	0	0	0	0	0
6	0	1	0	0	0	0	0	0	0	0	0	0	0	0
8	1	0	0	1	1	1	0	1	0	1	−1	−1	0	0
9	0	0	0	0	1	0	0	0	1	0	1	1	1	1
10	0	0	0	0	0	0	0	0	0	0	0	0	1	0
11	0	0	0	0	0	0	0	0	0	0	1	1	0	0
13	0	0	0	0	0	0	1	0	0	0	0	0	0	0
14	0	0	0	0	0	0	0	0	0	0	0	0	1	1

KEY

1 Richardson spermophile (ground squirrel)
2 marsh hawk, coyote, red-tailed hawk, weasel
3 badger
4 vole (*Microtus*)
5 13-striped spermophile (ground squirrel)
6 pocket gopher (*Thomomys*)
7 great horned owl
8 *Agropyron, Stipa, Helianthus*
9 insects in herb and surface stratum, Diptera, Hemiptera, grasshoppers, etc.
10 spiders
11 insects in soil stratum, wire worms, cutworms, white grubs, etc.
12 meadow lark, chipping sparrow, clay-colored sparrow, vesper sparrow, horned lark, upland plover
13 crow
14 frog
15 garter snake

1.2   Bird, 1930   393   Willow forest, Canada   Community

FOOD WEB MATRIX

	2	3	4	7	8	9	10	12
1	1	0	0	0	0	−1	0	0
5	1	0	0	0	1	−1	0	0
6	1	0	0	0	0	1	0	0
7	0	1	1	0	0	0	−1	0
8	0	−1	0	1	0	0	1	0
9	0	0	1	1	0	0	1	0
10	0	0	0	0	0	0	1	1
11	0	0	0	0	0	0	1	0

KEY

1 *Salix discolor*
2 *Galerucella decora*
3 redwinged blackbird, bronze grackle, song sparrow
4 Maryland yellowthroat, yellow warbler, song sparrow
5 *Salix petiolaris*
6 *Salix longifolia*
7 spiders
8 insects, *Pontania petiolaridis*, Collembola
9 insects, *Disyonicha quinquevitata*, Collembola
10 *Rana pipiens*
11 snails
12 garter snake

1.3   Bird, 1930   406   Aspen forest, Canada   Community

FOOD WEB MATRIX

	1	2	3	4	5	6	7	9	10	11	12	13	15	16	17	18	19	21	22	23	24	25
1	0	0	0	0	0	0	0	0	0	1	0	0	0	0	0	0	0	0	0	0	0	0
4	1	0	0	0	0	0	0	0	0	0	0	0	0	0	0	0	0	0	0	0	0	0
5	1	0	0	1	0	0	0	0	0	0	0	0	0	1	0	0	0	0	0	0	0	1
6	0	0	1	0	0	0	0	0	0	0	0	0	0	0	0	0	0	0	0	0	0	0
7	0	0	0	0	0	0	0	0	0	0	0	0	0	0	0	0	0	0	0	0	0	0
8	0	1	0	0	1	1	1	0	1	0	0	1	1	0	0	0	0	0	0	0	1	0
10	0	0	0	0	0	0	0	0	0	0	1	0	0	0	0	0	0	0	0	0	0	0
12	0	0	0	0	0	0	0	0	0	0	0	1*	0	0	0	0	0	0	0	0	0	0
13	0	0	0	0	0	0	0	1	0	0	0	0	0	0	0	0	0	0	0	0	0	0
14	0	0	0	0	0	0	0	0	0	0	1	0	0	0	0	0	0	0	0	0	0	0
15	0	0	0	0	0	0	0	0	0	0	1	0	0	0	0	0	0	0	0	0	0	0
16	0	0	0	0	0	0	0	0	0	1	0	0	0	0	0	0	0	0	0	0	0	0
18	0	0	0	0	0	0	0	1	0	0	1	0	0	0	1	0	0	0	0	0	0	0
19	0	0	0	0	0	0	0	0	0	1	0	0	0	0	0	0	0	0	0	0	0	0
20	0	1	0	0	0	1	0	0	0	0	0	1	1	0	0	1	0	1	1	0	0	0
21	0	0	0	0	0	0	0	0	0	0	1	0	0	0	0	0	0	0	0	0	0	0
22	0	0	0	0	0	0	0	0	0	0	0	0	0	1	0	0	1	0	0	1	0	0
24	0	0	0	0	0	0	0	0	0	0	0	0	0	0	0	0	0	0	0	0	0	1
25	0	0	0	0	0	0	0	0	0	0	0	0	0	0	0	0	0	0	0	0	0	1

KEY

1 Baltimore oriole, chickadee, least flycatcher, warbling vireo, rosebreasted grosbeak, willow thrush
2 canker, fomes
3 hairy and downy woodpeckers
4 spiders (mature forest)
5 insects (mature forest)
6 *Dicerca, Saperda*
7 red squirrel
8 *Populus, Cornus, Corylus, Pyrola, Aralia*
9 goshawk
10 redbacked vole (*Evolomys*)
11 Cooper's and sharpshinned hawks
12 great horned owl
13 ruffed grouse
14 flicker
15 crow
16 house wren
17 ticks
18 snowshoe rabbit
19 red-eyed vireo, yellow warbler, gold finch, catbird, brown thrasher, towhee, robin
20 *Populus, Symphoricarpos, Corylus, Prunus, Amelanchier*
21 redbacked vole, Franklin ground squirrel
22 insects (forest edge)
23 spiders (forest edge)
24 snails
25 frogs

* It is unlikely that the ruffed grouse preys on the great horned owl, as shown here, in the original, and in food web 1.4. The direction of predation probably should be reversed.

FOOD WEB MATRIX

	1	2	3	4	5	6	8	9	10	11	12	13	14	15	16	18	20	22	23	24	25	28	29	30	33
1	0	0	0	0	0	0	0	1	0	0	0	0	0	1	0	0	0	0	0	0	0	0	0	0	0
4	0	0	1	0	0	0	0	0	0	0	0	0	0	0	0	0	0	0	0	0	0	0	0	0	0
5	1	0	0	0	0	0	1	0	−1	0	0	0	0	0	0	1	0	0	1	1	0	0	0	0	0
6	1	0	0	0	0	0	0	0	0	0	0	0	0	0	0	1	0	0	0	1	0	0	0	0	0
7	0	1	0	1	1	1	0	0	1	1	1	0	1	0	0	0	1	0	0	0	0	0	0	0	0
9	0	0	0	0	0	0	0	0	0	0	0	0	1	0	0	0	0	0	0	0	0	0	0	0	0
10	0	0	0	0	0	0	0	0	0	1	1	0	0	0	0	0	0	0	0	0	0	0	0	0	0
13	0	0	0	0	0	0	0	0	0	1	1	0	1	0	0	0	0	0	0	0	0	1	0	0	0
14	0	0	0	0	0	0	0	0	1	0	0	0	0	0	0	0	0	0	0	0	0	0	0	0	0
15	0	0	0	0	0	0	0	0	0	0	0	0	1	0	0	0	0	0	0	0	0	0	0	0	0
16	0	0	0	0	0	0	1	0	0	0	0	0	0	0	0	0	0	0	0	0	0	0	0	0	0
17	0	0	0	0	1	0	0	0	0	0	0	0	1	0	0	0	1	0	0	0	0	0	0	0	0
19	0	0	0	0	0	0	0	0	0	0	0	0	0	0	0	1	0	0	0	0	0	0	0	0	0
20	0	0	0	0	0	0	0	0	0	0	0	0	1	0	0	0	0	0	0	0	0	0	0	0	0
21	0	0	0	0	1	0	0	0	0	0	0	0	1	0	0	0	1	0	0	0	0	0	0	0	0
22	0	0	0	0	0	0	1	0	0	0	0	0	0	0	0	0	0	0	0	0	0	0	0	0	0
23	0	0	0	0	0	0	0	0	0	0	0	0	0	0	0	0	0	0	1	1	1	0	0	0	0
24	0	0	0	0	0	0	0	0	0	0	0	0	0	0	0	0	0	0	0	0	1	0	0	0	0
25	0	0	0	0	0	0	0	0	0	1	0	0	0	0	0	0	0	0	0	0	0	0	0	0	0
26	0	0	0	0	0	0	0	0	0	0	0	0	1	0	0	0	0	0	0	0	0	0	0	0	0
27	0	0	0	0	0	0	0	0	0	1	0	0	0	0	0	0	0	0	0	0	0	0	0	0	0
28	0	0	0	0	0	0	0	0	0	1	0	0	0	0	0	0	0	0	0	0	0	0	0	0	0
29	0	0	0	0	0	0	0	0	0	0	0	0	1	0	0	0	0	0	0	0	0	1	0	1	0
30	0	0	0	0	0	0	0	0	0	0	0	0	1	0	0	0	0	0	0	0	0	1	0	0	0
31	0	0	0	0	0	0	0	−1	0	1	0	0	0	0	0	0	0	0	0	0	0	0	1	1	1
32	0	0	0	0	0	0	0	0	0	0	0	0	1	0	0	0	0	0	0	0	0	0	0	0	0
33	0	0	0	0	0	0	0	1	0	0	0	0	0	0	0	0	0	0	0	1	0	0	0	0	0
34	0	0	0	0	0	0	0	0	0	0	0	0	0	1	0	0	0	0	0	0	0	0	0	0	0

KEY

1 Baltimore oriole, chickadee, least flycatcher, rosebreasted grosbeak, willow thrush
2 canker, fomes
3 hairy and downy woodpeckers
4 *Dicerca, Saperda*
5 insects
6 spiders
7 *Populus, Cornus, Corylus, Pyrola, Aralia*
8 Cooper's and sharpshinned hawks
9 crow
10 ruffed grouse
11 man
12 goshawk
13 rabbit
14 great horned owl
15 flicker
16 Maryland yellowthroat, yellow warbler, song sparrow
17 *Salix longifolia*
18 *Galerucella decora*
19 *Salix discolor*
20 red squirrel
21 *Salix petiolaris*
22 yellow warbler, redwinged blackbird, bronze grackle
23 *Rana pipiens*
24 garter snake
25 fish
26 coots
27 ducks
28 coyote, weasel, skunk
29 prairie vole
30 pocket gophers, ground squirrels
31 *Agrostis, Agropyron, Stipa, Helianthus*
32 mice
33 cutworms, grasshoppers, clickbeetles
34 ants

2   Clarke et al., 1967   1384   Sand bottom   Community

FOOD WEB MATRIX

	1	2	3	4	5	6	7	8	9	10	12	13	14	15	20	21	22	24
9	0	1	1	1	1	-1	0	0	0	0	0	0	0	0	0	0	0	0
10	0	0	1	0	1	-1	0	0	0	0	0	0	0	0	0	0	0	0
11	0	0	0	0	0	0	-1	-1	0	0	0	0	0	0	0	0	0	0
16	0	0	0	0	0	0	0	0	0	0	0	0	0	1	0	0	0	0
17	0	0	0	0	0	0	0	0	0	0	0	0	1	1	0	0	0	0
18	0	0	0	0	0	0	0	0	0	0	1	-1	1	1	0	0	0	0
19	0	0	0	0	0	0	0	0	0	0	1	0	1	0	0	0	0	0
20	0	1	1	1	0	0	0	0	0	0	0	0	0	0	0	0	0	0
21	-1	1	1	1	0	0	0	0	0	0	0	0	0	0	0	0	0	0
22	-1	1	1	1	1	0	0	0	1	0	0	0	0	0	1	0	0	0
23	0	1	0	0	0	0	0	0	0	-1	0	0	0	0	1	1	-1	1

KEY

1 Pacific bonito
2 vermilion rockfish
3 California scorpion fish
4 California sea lion
5 cabezon
6 sand bass
7 Pacific angel shark
8 California halibut
9 squid
10 octopus
11 small fishes and invertebrates
12 Pacific sand dab
13 pink sea perch
14 hornyhead turbot
15 longfin sand dab
16 ophiuroids
17 polychaetes
18 benthic crustacea
19 hypoplanktonic crustacea
20 white croaker
21 jack mackerel
22 northern anchovy
23 zooplankton
24 blacksmith, black perch, rubber lip sea perch, sharpnose sea perch, shiner perch

4   Hairston, 1949   68   Salamanders, Appalachians   Sink

FOOD WEB MATRIX

	1	2	3	4
5	1	0	0	0
6	2	0	0	0
7	2	0	0	0
8	4	0	0	0
10	0	1	0	0
11	0	1	0	0
12	0	1	0	0
13	3	0	0	0
14	1	0	0	0
15	1	2	0	0
16	2	1	0	0
17	1	1	0	3
18	2	3	0	1
19	3	0	1	2
20	1	0	0	1
21	2	0	0	1
22	0	1	0	0
23	0	1	0	0
24	0	1	0	0
25	0	1	0	0
26	0	1	3	0
27	0	1	1	4
28	0	0	1	0
29	0	0	1	0
30	0	0	1	0
31	0	0	2	0
32	0	0	1	0
33	0	0	1	0
34	0	0	1	0
35	0	0	1	0
36	0	0	1	0
37	0	0	1	0
38	0	0	3	9
39	0	0	1	4
40	0	0	0	1
41	0	0	0	2
42	0	0	0	3
43	0	0	0	1
44	0	0	0	2
45	0	0	0	3

KEY

1 *Desmognathus quadramaculatus*
2 *Desmognathus monticola*
3 *Desmognathus carolinensis*
4 *Desmognathus wrighti*
5 Plecoptera (naiad)
6 Diptera—Dixidae (larvae)
7 Dixidae or Chironomidae (larvae)
8 Dixidae or Chironomidae (adults)
10 Bembidiini
11 Staphylinidae—*Stenus* sp.
12 Chrysomelidae—*Donacia* sp.
13 Homoptera—Cercopidae
14 Hymenoptera—unidentified
15 Formicidae
16 Diptera—Tipulidae (adults)
17 Collembola
18 Araneida
19 Hymenoptera—parasitic wasp
20 Collembola—Entomobryidae
21 Lepidoptera—Heterocera
22 Diplopoda—Julidae
23 Polydesmidae
24 Coleoptera—Photuridae *Lucidota* sp.
25 Hymenoptera—Formicidae *Camponotus* sp. (queens)
26 Homoptera—Cicadellidae
27 Pseudoscorpionida
28 Oligochaeta—Lumbricidae
29 Pulmonata—Zonitidae *Zonitoides arboreus*
30 Orthoptera—Locustidae
31 Lepidoptera—Rhopalocera (larva)
32 Coleoptera—Staphylinidae *Atheta* sp.
33 Cisidae
34 Cantharidae—*Cantharis* sp.
35 Diptera—Tipulidae (larvae)
36 Cecidomyidae
37 Hymenoptera—Cynipidae
38 Acarina—Orabatidae
39 Parasitidae
40 Thysanoptera
41 Collembola—Poduridae
42 Sminthuridae
43 Diptera—Mycetophilidae
44 Acarina—Trombidiidae
45 Hoplodermatidae

FOOD WEB MATRIX

	1	2	3	4	5	6	7	8	9	10	14	18	20	21	23	24	25	29
1	0	0	-2	0	-1	-1	-1	-1	0	0	0	0	0	0	0	0	0	0
2	0	0	-2	0	-1	-1	0	0	0	0	0	0	0	0	0	0	0	0
8	0	0	0	1	-1	-1	0	-1	0	0	0	0	0	0	0	0	0	0
9	0	0	0	1	0	0	0	0	0	0	0	0	0	0	0	0	0	0
10	0	0	1	1	0	0	0	0	0	0	0	0	0	0	0	0	0	0
11	0	0	0	1	0	0	0	0	0	0	0	0	0	0	0	0	0	0
12	-2	0	1	1	0	0	0	0	0	0	0	0	0	0	0	0	0	0
13	0	0	1	0	0	0	0	0	0	0	0	0	0	0	0	0	0	0
14	0	0	1	1	-1	-1	0	0	0	0	0	0	0	0	0	0	0	0
15	0	0	0	1	0	0	0	0	0	0	0	0	0	0	0	0	0	0
16	0	0	0	1	0	0	0	0	0	0	0	0	0	0	0	0	0	0
17	0	0	1	1	0	0	0	0	0	0	0	0	0	0	0	0	0	0
18	1	0	0	0	0	0	0	0	0	0	0	0	0	0	0	0	0	0
19	1	0	1	0	0	0	0	0	0	0	0	0	0	0	0	0	0	0
21	1	1	1	0	-1	0	0	-1	0	1	0	0	0	0	0	0	0	0
22	0	0	1	0	-1	0	0	0	0	0	0	0	0	0	0	0	0	0
23	0	1	1	1	-1	0	0	-1	0	1	0	0	0	0	0	0	0	0
24	0	0	1	0	0	0	0	0	0	0	0	0	0	0	0	0	0	0
25	0	0	0	1	0	0	0	-1	0	0	0	0	0	0	0	0	0	0
26	0	0	0	1	0	0	0	0	0	0	0	0	0	0	0	0	0	0
27	0	0	0	1	0	0	0	0	0	0	0	0	0	0	0	0	0	0
28	1	0	0	0	0	0	0	0	0	0	0	0	0	0	0	0	0	0
30	0	0	0	0	0	0	0	0	-1	0	0	0	0	0	0	0	0	0
31	0	0	0	0	0	0	0	0	-1	0	0	0	0	0	0	0	0	0
32	1	0	0	0	0	0	0	0	0	0	0	0	0	0	0	0	0	0
33	1	0	0	0	0	0	0	0	0	0	0	0	0	0	0	-1	-1	0
34	0	0	0	0	0	0	0	0	0	0	0	0	-1	-1	0	0	0	-1
35	0	0	0	0	0	0	0	0	0	0	0	0	-1	-1	-1	-1	-1	0
36	0	0	0	0	0	0	0	0	0	0	0	0	0	0	0	0	0	-1
37	1	0	0	0	0	0	0	0	-1	0	-1	-1	-1	-1	-1	0	-1	-1
38	0	0	0	0	0	0	0	0	0	0	-1	-1	0	0	0	-1	-1	0
39	0	0	0	0	0	0	0	0	0	0	-1	-1	0	0	0	0	-1	-1
40	0	0	0	0	0	0	0	0	0	0	0	0	0	0	0	0	-1	0
41	0	0	0	0	0	0	0	0	0	0	0	0	0	0	0	0	-1	0
42	0	0	0	0	0	0	0	0	0	0	0	0	0	0	-1	0	-1	0
43	0	0	0	0	0	0	0	0	0	0	0	0	0	0	-1	0	0	0

KEY

1 7-12mm young herring
2 12-42mm young herring
3 42-130mm young herring
4 adult herring
5 medusae
6 *Pleurobrachia*
7 *Tomopteris*
8 *Sagitta*
9 *Limacina*
10 *Ammodytes* juv.
11 *Oikopleura*
12 *Cypris balanus* larvae
13 Mysidae
14 Decapod larvae
15 *Nyctiphanes*
16 Hyperiid amphipods
17 *Apherusa*
18 Larval mollusca
19 Harpacticid
20 *Paracalanus*
21 *Pseudocalanus*
22 *Acartia*

23 *Temora*
24 *Centropages*
25 *Calanus*
26 *Evadne*
27 *Podon*
28 *Tintinnopsis*
29 *Noctiluca*
30 Radiolaria
31 Foraminifera
32 *Prorocentrum*
33 *Peridinium*
34 *Melosira*
35 *Thalassiosira*
36 *Guinardia*
37 *Coscinodiscus*
38 *Phaetocystis*
39 *Rhizosolenia*
40 *Biddulphia*
41 *Chaetoceras*
42 *Nitzschia*
43 *Navicula*

FOOD WEB MATRIX

	6	7	8	9	12	13	14	15	16	17	18	19	20	21	22	23	24	25	26	27	28	29	30	31	32	33	34	35	36
1	0	0	0	0	1	1	1	1	1	1	1	1	0	0	1	0	0	0	0	0	0	0	0	0	0	0	0	0	0
2	0	0	0	0	0	0	0	0	1	0	0	0	0	0	0	0	0	0	0	0	0	0	0	0	0	0	0	0	0
3	0	0	0	0	0	0	0	0	-1	-1	0	0	0	0	0	0	0	1	0	0	0	0	0	0	0	0	0	0	0
4	0	0	0	0	0	0	0	0	0	0	0	0	-1	-1	0	0	0	-1	0	0	0	0	0	0	0	0	0	0	0
5	0	-1	-1	-1	0	0	0	0	0	0	0	0	0	0	0	-1	0	0	1	0	0	0	0	0	0	0	0	0	0
6	0	0	0	0	0	0	0	0	0	0	0	0	0	0	0	-1	1	1	-1	1	1	1	1	-1	1	1	1	1	1
7	1	0	0	0	0	0	0	0	0	0	0	0	0	0	0	1	0	0	0	0	0	0	0	0	0	0	0	0	0
8	1	0	0	0	0	0	0	0	0	0	0	0	0	0	0	1	0	0	0	0	0	0	0	0	0	0	0	0	0
9	1	0	0	0	0	0	0	0	0	0	0	0	0	0	0	1	0	0	0	0	0	0	0	0	0	0	0	0	0
10	1	0	0	0	0	0	0	0	0	0	0	0	0	0	0	0	0	0	0	0	0	0	0	0	0	0	0	0	0
11	1	0	0	0	0	0	0	0	0	0	0	0	0	0	0	0	0	0	0	0	0	0	0	0	0	0	0	0	0
12	1	0	0	0	0	0	0	0	0	0	0	0	0	0	0	0	0	0	0	0	0	0	0	0	0	0	0	0	0
13	1	0	0	0	0	0	0	0	0	0	0	0	0	0	0	0	0	0	0	0	0	0	0	0	0	0	0	0	0
14	1	0	0	0	0	0	0	0	0	0	0	0	0	0	0	0	0	0	0	0	0	0	0	0	0	0	0	0	0
15	0	0	0	0	0	0	0	0	0	0	0	0	0	0	0	0	0	0	0	0	0	0	0	0	0	0	0	0	0
16	1	0	0	0	0	0	0	0	0	0	0	0	0	0	0	1	0	0	0	0	0	0	0	0	0	0	0	0	0
18	0	0	0	0	0	0	0	0	0	0	0	0	0	0	0	0	0	0	0	0	0	0	0	0	0	0	0	0	0
19	0	0	0	0	0	0	0	0	0	0	0	0	0	0	0	0	0	0	0	0	0	0	0	0	0	0	0	0	0
20	0	0	0	0	0	0	0	0	0	0	0	0	0	0	0	0	0	0	0	0	0	0	0	0	0	0	0	0	0
21	0	0	0	0	0	0	0	0	0	0	0	0	0	0	0	0	0	0	0	0	0	0	0	0	0	0	0	0	0
22	0	0	0	0	0	0	0	0	0	0	0	0	0	0	0	1	0	0	0	0	0	0	0	0	0	0	0	0	0
23	0	0	0	0	0	0	0	0	0	0	0	0	0	0	0	0	0	0	0	0	0	0	0	0	0	0	0	1	1
25	0	0	0	0	0	0	0	0	0	0	0	0	0	0	0	1	0	0	0	0	0	0	0	0	0	0	0	0	0
26	0	0	0	0	0	0	0	0	0	0	0	0	0	0	0	1	0	0	0	0	0	0	0	0	0	0	0	0	0
27	0	0	0	0	0	0	0	0	0	0	0	0	0	0	0	1	0	0	0	0	0	0	0	0	0	0	0	0	0
28	0	0	0	0	0	0	0	0	0	0	0	0	0	0	0	1	0	0	0	0	0	0	0	0	0	0	0	0	0
29	0	0	0	0	0	0	0	0	0	0	0	0	0	0	0	1	0	0	0	0	0	0	0	0	0	0	0	0	0
35	0	0	0	0	0	0	0	0	0	0	0	0	0	0	0	0	0	0	0	0	0	0	0	0	0	0	0	0	1
37	0	0	0	0	0	0	0	0	0	0	0	0	0	0	0	0	0	0	0	0	0	0	0	0	0	0	0	0	0
38	0	0	0	0	0	0	0	0	0	0	0	0	0	0	0	0	0	0	0	0	0	0	0	0	0	0	0	0	0
42	0	0	0	0	0	0	0	0	0	0	0	0	0	0	0	0	0	0	0	0	0	0	0	0	0	0	0	0	0
43	0	0	0	0	0	0	0	0	0	0	0	0	0	0	0	0	0	0	0	0	0	0	0	0	0	0	0	0	0
44	0	0	0	0	0	0	0	0	0	0	0	0	0	0	0	0	0	0	0	0	0	0	0	0	0	0	0	0	0
45	0	0	0	0	0	0	0	0	0	0	0	0	0	0	0	0	0	0	0	0	0	0	0	0	0	0	0	0	0
49	0	0	0	0	0	0	0	0	0	0	0	0	0	0	0	0	0	0	0	0	0	0	0	0	0	0	0	0	0
50	0	0	0	0	0	0	0	0	0	0	0	0	0	0	0	0	0	0	0	0	0	0	0	0	0	0	0	0	0
51	0	0	0	0	0	0	0	0	0	0	0	0	0	0	0	0	0	0	0	0	0	0	0	0	0	0	0	0	0
56	0	0	0	0	0	0	0	0	0	0	0	0	0	0	0	0	0	0	0	0	0	0	0	0	0	0	0	0	0
57	0	0	0	0	0	0	0	0	0	0	0	0	0	0	0	0	0	0	0	0	0	0	0	0	0	0	0	0	0
60	1	0	0	0	0	0	0	0	0	0	0	0	0	0	0	0	0	0	0	0	0	0	0	0	0	0	0	0	0
61	1	0	0	0	0	0	0	0	0	0	0	0	0	0	0	0	0	0	0	0	0	0	0	0	0	0	0	0	0
62	1	0	0	0	0	0	0	0	0	0	0	0	0	0	0	0	0	0	0	0	0	0	0	0	0	0	0	0	0

KEY

1 plancton marino y detritus
2 diatomea de la zona de la rompiente (en grano de arena)
3 diatomeas de la pelicula superficial en la zona intercotidal
4 fauna del agua subterranea: como ejemplo un foraminifero frecuente
5 *Macrocystis pyrifera humboldtii*
6 *Phalacrocorax bougainvillii* y sula variegata
7 *Tetracha chilensis*
8 *Dysderdus* spec.
9 *Actinote* spec.
10 *Hepatus chiliensis*
11 *Pseudocorystes sicarius*
12 *Donax paytensis*
13 *Mesodesma donacium*
14 *Tivela planulata*
15 *Olivella columellaris*
16 *Emerita analoga*
17 *Emerita emerita*
18 *Blepharipoda occidentalis*
19 *Lepidopa chilensis*
20 *Callianassa* spec.
21 *Austromenidia regia* y *Engraulis ringens*
22 Talitrido
23 *Ocypode gaudichaudii*
24 *Coenobita compressus*
25 moscas tipicas de las algas
26 moscas de las madrigueras de ocypode
27 *Calliphora* spec.
28 *Dermestes maculatus*
29 *Phaleria* spec.
30 Staphylinido

	37	38	39	40	41	42	43	44	45	46	47	48	49	50	51	52	53	54	55	56	57	58	59	60	61	62	63	64	65
1	0	0	0	0	0	0	0	0	0	0	0	0	0	0	0	0	0	0	0	0	0	0	0	0	0	0	0	0	0
2	0	0	0	0	0	0	0	0	0	0	0	0	0	0	0	0	0	0	0	0	0	0	0	0	0	0	0	0	0
3	0	0	0	0	0	0	0	0	0	0	0	0	0	0	0	0	0	0	0	0	0	0	0	0	0	0	0	0	0
4	0	0	0	0	0	0	0	0	0	0	0	0	0	0	0	0	0	0	0	0	0	0	0	0	0	0	0	0	0
5	0	0	0	0	0	0	0	0	0	0	0	0	0	0	0	0	0	0	0	0	0	0	0	0	0	0	0	0	0
6	0	0	0	0	0	0	0	0	0	0	0	0	0	0	0	0	0	0	0	0	0	0	0	0	0	0	0	0	0
7	0	0	0	0	0	0	0	0	1	0	0	0	0	1	1	1	1	1	0	1	1	1	0	0	0	0	0	0	0
8	0	0	0	0	0	0	0	0	1	0	0	0	0	1	1	1	1	1	0	1	1	1	0	0	0	0	0	0	0
9	0	0	0	0	0	0	0	0	1	0	0	0	0	1	1	1	1	1	0	1	1	1	0	0	0	0	0	0	0
10	0	0	0	0	0	0	0	0	0	0	0	0	0	0	0	0	0	0	0	0	0	0	0	0	0	0	0	0	0
11	0	0	0	0	0	0	0	0	0	0	0	0	0	0	0	0	0	0	0	0	0	0	0	0	0	0	0	0	0
12	0	0	0	0	0	0	0	0	0	0	0	0	0	0	0	0	0	0	0	0	0	0	0	0	0	0	0	0	0
13	0	0	0	0	0	0	0	0	0	0	0	0	0	0	0	0	0	0	0	0	0	0	0	0	0	0	0	0	0
14	0	0	0	0	0	0	0	0	0	0	0	0	0	0	0	0	0	0	0	0	0	0	0	0	0	0	0	0	0
15	0	0	0	0	0	0	0	0	1	0	0	0	0	0	0	0	0	0	0	1	1	0	0	0	0	0	0	0	0
16	1	1	1	1	1	1	1	1	1	0	0	0	1	0	1	0	0	0	0	1	1	0	1	0	0	0	0	0	0
18	0	0	0	0	1	0	0	0	0	0	0	0	0	0	0	0	0	0	0	0	0	0	0	0	0	0	0	0	0
19	0	0	0	1	1	0	0	0	0	0	0	0	0	0	0	0	0	0	0	0	0	0	0	0	0	0	0	0	0
20	0	0	1	0	1	1	1	1	0	0	0	0	0	0	0	0	0	0	0	0	0	0	0	0	0	0	0	0	1
21	0	0	0	0	0	1	1	1	0	0	0	0	0	0	0	0	0	0	0	0	0	1	1	1	1	1	0	0	0
22	0	0	0	0	0	0	0	0	1	0	0	0	0	1	1	1	1	1	0	1	1	1	0	0	0	0	0	0	0
23	0	1	0	0	0	0	0	0	0	1	1	-1	1	1	0	1	1	0	0	0	1	1	0	0	0	0	0	0	0
25	0	0	0	0	0	0	0	0	1	0	0	0	0	1	1	1	1	1	0	1	1	1	0	0	0	0	0	0	0
26	0	0	0	0	0	0	0	0	1	0	0	0	0	1	1	1	1	1	0	1	1	1	0	0	0	0	0	0	0
27	0	0	0	0	0	0	0	0	1	0	0	0	0	1	1	1	1	1	0	1	1	1	0	0	0	0	0	0	0
28	0	0	0	0	0	0	0	0	1	0	0	0	0	1	1	1	1	1	0	1	1	1	0	0	0	0	0	0	0
29	0	0	0	0	0	0	0	0	1	0	0	0	0	1	1	1	1	1	0	1	1	1	0	0	0	0	0	0	0
35	0	0	0	0	0	0	0	0	0	0	0	0	0	0	0	0	0	0	0	0	0	0	0	0	0	0	0	0	0
37	0	0	0	0	0	0	0	0	0	0	0	0	0	0	0	0	0	0	0	0	0	0	0	0	0	0	0	1	0
38	0	0	0	0	0	0	0	0	0	0	0	0	0	0	0	0	0	0	0	0	0	0	0	0	0	0	0	1	0
42	0	0	0	0	0	0	0	0	0	0	0	0	0	0	0	0	0	0	0	0	0	0	0	0	0	0	1	0	0
43	0	0	0	0	0	0	0	0	0	0	0	0	0	0	0	0	0	0	0	0	0	0	0	0	0	0	1	0	0
44	0	0	0	0	0	0	0	0	0	0	0	0	0	0	0	0	0	0	0	0	0	0	0	0	0	0	1	0	0
45	0	0	0	0	0	0	0	0	0	0	0	0	0	0	0	0	0	0	0	0	0	0	0	0	0	0	0	1	0
49	0	0	0	0	0	0	0	0	0	0	0	0	0	0	0	0	0	0	0	0	0	0	0	0	0	0	0	1	0
50	0	0	0	0	0	0	0	0	0	0	0	0	0	0	0	0	0	0	0	1	0	0	0	0	0	0	0	0	0
51	0	0	0	0	0	0	0	0	0	0	0	0	0	0	0	0	0	0	0	1	0	0	0	0	0	0	0	0	0
56	0	0	0	0	0	0	0	0	0	0	0	0	0	0	0	0	0	0	0	0	0	0	0	0	0	0	0	1	0
57	0	0	0	0	0	0	0	0	0	0	0	0	0	0	0	0	0	0	0	0	0	0	0	0	0	0	0	1	0
60	0	0	0	0	0	0	0	0	0	0	0	0	0	0	0	0	0	0	0	0	0	0	0	0	0	0	0	0	0
61	0	0	0	0	0	0	0	0	0	0	0	0	0	0	0	0	0	0	0	0	0	0	0	0	0	0	0	0	0
62	0	0	0	0	0	0	0	0	0	0	0	0	0	0	0	0	0	0	0	0	0	0	0	0	0	0	0	0	0

KEY

31 *Coragyps atratus*
32 *Cathartes aura jota*
33 *Vultur gryphus*
34 *Caracara plancus cheriway*
35 *Rattus rattus alexandrinus*
36 *Dusicyon sechurae*
37 *Larus modestus*
38 *Crocethia alba*
39 *Aetobatus peruvianus*
40 *Rhinobatos planiceps*
41 *Mustelus* spec.
42 *Sciaena gilberti*
43 *Umbrina xanti*
44 *Polynemus* spec.
45 los limicolae mas grandes
46 *Conepatus* spec.
47 *Leucopternis schistacea plumbea*
48 *Larus belcheri*
49 *Larus pipixcan*

50 *Brachystosternus ehrenbergi*
51 *Tropidurus peruvianus*
52 Golondrinas como: *Hirundo rustica erythrogaster*
53 Murcielagos (diversas especies)
54 *Chordeiles acutipennis* ssp.
55 *Falco sparverius peruvianus*
56 diversos Charadriidae
57 *Arenaria interpres morinella* y *Actitis macularia*
58 *Geositta peruviana paytae*
59 *Paralichthys adspersus*
60 *Pelecanus occidentalis thagus*
61 *Phalacrocorax bougainvillii*
62 *Sula variegata*
63 *Pandion haliaetus carolinensis*
64 *Falco peregrinus anatum*
65 *Clausidium* spec.

8.1 Kohn, 1959 70 *Conus*, Hawaiian marine benches Sink

FOOD WEB MATRIX

	1	2	3	4	5	6	7	8	9
10	30	0	0	0	0	0	0	0	0
11	32	20	115	0	13	0	0	0	0
12	10	3	0	45	0	1	0	0	0
13	4	0	0	0	0	0	0	0	0
14	1	1	0	0	0	0	0	0	0
15	12	28	0	0	0	0	0	1	0
16	0	0	7	8	0	0	0	0	0
17	2	2	0	0	12	0	0	0	0
18	0	0	0	0	1	0	0	0	0
19	1	2	0	0	0	0	0	0	0
20	1	3	0	0	0	0	0	0	0
21	33	7	0	0	0	0	0	0	1
22	1	1	1	0	0	0	0	0	0
23	14	11	0	0	0	0	0	0	0
24	0	3	0	0	0	0	0	0	0
25	0	0	0	0	0	4	2	0	0
26	0	0	0	0	0	0	3	0	0
27	0	0	0	0	0	1	0	0	0
28	0	0	0	0	0	0	1	0	0
29	6	4	1	0	0	2	1	0	0
30	3	2	0	0	0	0	0	0	0

KEY

1 *Conus sponsalis*
2 *Conus abbreviatus*
3 *Conus ebraeus*
4 *Conus chaldaeus*
5 *Conus rattus*
6 *Conus lividus*
7 *Conus flavidus*
8 *Conus miles*
9 *Conus distans*
10 *Nereis jacksoni*
11 *Perinereis helleri*
12 *Platynereis dumerillii*
13 Nereid sp. 350
14 unidentified Nereidae
15 *Lysidice collaris*

16 *Palola siciliensis*
17 *Eunice antennata*
18 *Eunice (Nicidion) cariboea*
19 *Eunice filamentosa*
20 *Marphysa sanguinea*
21 *Eunice afra*
22 unidentified Eunicidae
23 *Lumbrinereis sarsi*
24 *Arabella iricolor*
25 *Nicolea gracilibranchus*
26 Terebellid sp. 837
27 *Cirriformia semicincta*
28 Polydorid sp. 1500
29 unidentified annelids
30 Onuphid

8.2   Kohn, 1959   74-75   *Conus*, Hawaiian reef, benches and deep water   Sink

FOOD WEB MATRIX

	1	2	3	4	5	6	7	8	9	10	11	12	13
14	34	0	0	0	0	0	0	0	0	0	0	0	0
15	35	20	136	5	0	21	0	0	0	0	0	0	0
16	13	3	0	98	0	1	0	0	0	0	0	0	12
17	5	0	0	0	0	0	0	0	0	0	0	0	0
18	2	1	0	0	0	0	0	0	0	0	0	0	0
19	18	38	2	0	16	0	0	0	0	0	0	0	0
20	0	1	44	14	0	0	0	0	0	0	0	0	0
21	3	22	0	0	1	29	1	9	0	0	0	0	0
22	39	14	1	0	0	2	0	0	0	0	0	0	0
23	36	10	0	0	0	9	13	0	1	0	0	0	0
24	1	6	0	0	0	0	0	6	0	4	0	0	0
25	1	3	0	0	0	0	0	0	0	0	0	0	0
26	1	2	1	1	0	0	0	0	0	0	0	0	0
27	17	11	0	0	0	0	0	0	0	0	0	0	0
28	0	9	0	0	0	0	0	0	0	0	0	0	0
29	0	0	0	0	0	0	0	0	0	11	0	0	0
30	7	4	2	0	0	0	0	0	0	0	0	0	0
31	0	0	0	0	0	0	0	0	0	0	2	24	0
32	0	0	0	0	0	0	0	0	0	0	0	11	0
33	0	0	0	0	0	0	0	0	0	0	0	22	1
34	0	0	0	0	0	0	0	0	0	0	0	7	1
35	0	0	0	0	0	0	0	0	0	0	0	34	11
36	0	0	0	0	0	0	0	0	0	0	0	0	2
37	0	0	0	0	0	0	0	0	0	0	0	5	1
38	0	0	0	0	0	0	0	0	0	0	0	2	11
39	0	0	0	0	0	0	0	0	0	0	0	0	2
40	0	0	0	0	0	0	0	0	0	0	0	0	3
41	0	0	0	0	0	0	0	0	0	0	0	5	3
42	0	0	0	0	0	0	0	0	0	0	2	0	0
43	0	0	0	0	0	0	0	0	0	0	0	1	33
44	0	0	0	0	0	0	0	0	0	0	0	4	32
45	0	0	0	0	0	0	0	0	0	0	0	0	1
46	3	2	0	0	0	0	0	0	0	0	0	0	0

KEY

1 *Conus sponsalis*
2 *Conus abbreviatus*
3 *Conus ebraeus*
4 *Conus chaldaeus*
5 *Conus miles*
6 *Conus rattus*
7 *Conus distans*
8 *Conus vexillum*
9 *Conus vitulinus*
10 *Conus imperialis*
11 *Conus pulicarius*
12 *Conus flavidus*
13 *Conus lividus*
14 *Nereis jacksoni*
15 *Perinereis helleri*
16 *Platynereis dumerilii*
17 *Nereid* sp. 350
18 unidentified Nereidae
19 *Lysidice collaris*
20 *Palola siciliensis*
21 *Eunice antennata*
22 *Eunice (N.) cariboea*
23 *Eunice afra*

24 *Marphysa sanguinea*
25 *Eunice filamentosa*
26 unidentified Eunicidae
27 *Lumbrinereis sarsi*
28 *Arabella iricolor*
29 *Eurythoe complanata*
30 unidentified annelids
31 *Capitellid* sp. 1040
32 *Thelepus setosus*
33 *Polycirrus* sp. 660
34 *Terebellid* sp. 837
35 *Nicolea gracilibranchus*
36 unidentified Terebellidae
37 *Polydorid* sp. 1500
38 *Cirriformia semicincta*
39 *Lygdamis nesiotes*
40 *Sabellastarte indica*
41 unidentified annelids
42 *Thalassema* sp.
43 *Ptychodera flava laysanica*
44 *P. flava laysanica* (tentative)
45 *Octopus* sp.
46 Onuphid

8.3   Kohn, 1959   74-75   *Conus*, Hawaiian subtidal reefs only   Sink

FOOD WEB MATRIX

	1	2	3	4	5	6	7	8	9	10	11	12	13
14	4	0	0	0	0	0	0	0	0	0	0	0	0
15	3	0	6	0	0	7	0	0	0	0	0	0	0
16	3	0	0	2	0	1	0	0	0	0	0	0	11
17	1	0	0	0	0	0	0	0	0	0	0	0	0
18	1	1	0	0	0	0	0	0	0	0	0	0	0
19	6	10	3	0	5	0	0	0	0	0	0	0	0
20	0	1	29	14	0	0	0	0	0	0	0	0	0
21	1	20	0	0	1	17	1	9	0	0	0	0	0
22	3	4	1	0	1	0	1	0	0	0	0	0	0
23	3	13	0	0	0	10	12	0	1	0	0	0	0
24	0	3	0	0	0	0	0	6	0	4	0	0	0
25	0	0	0	0	0	0	0	0	0	0	0	0	0
26	0	1	0	0	0	0	0	0	0	0	0	0	0
27	3	0	0	0	0	0	0	0	0	0	0	0	0
28	0	5	0	0	0	0	0	0	0	0	0	0	0
29	0	0	0	0	0	0	0	0	0	11	0	0	0
30	1	0	1	0	0	0	0	0	0	0	0	0	0
31	0	0	0	0	0	0	0	0	0	0	2	24	0
32	0	0	0	0	0	0	0	0	0	0	0	11	0
33	0	0	0	0	0	0	0	0	0	0	0	22	1
34	0	0	0	0	0	0	0	0	0	0	0	2	1
35	0	0	0	0	0	0	0	0	0	0	0	32	7
36	0	0	0	0	0	0	0	0	0	0	0	0	1
37	0	0	0	0	0	0	0	0	0	0	0	5	1
38	0	0	0	0	0	0	0	0	0	0	0	1	4
39	0	0	0	0	0	0	0	0	0	0	0	0	2
40	0	0	0	0	0	0	0	0	0	0	0	0	3
41	0	0	0	0	0	0	0	0	0	0	0	4	2
42	0	0	0	0	0	0	0	0	0	0	2	0	0
43	0	0	0	0	0	0	0	0	0	0	0	1	28
44	0	0	0	0	0	0	0	0	0	0	0	1	29
45	0	0	0	0	0	0	0	0	0	0	0	0	1

KEY

1 *Conus sponsalis*
2 *Conus abbreviatus*
3 *Conus ebraeus*
4 *Conus chaldaeus*
5 *Conus miles*
6 *Conus rattus*
7 *Conus distans*
8 *Conus vexillum*
9 *Conus vitulinus*
10 *Conus imperialis*
11 *Conus pulicarius*
12 *Conus flavidus*
13 *Conus lividus*
14 *Nereis jacksoni*
15 *Perinereis helleri*
16 *Platynereis dumerilii*
17 Nereid sp. 350
18 unidentified Nereidae
19 *Lysidice collaris*
20 *Palola siciliensis*
21 *Eunice antennata*
22 *Eunice (N.) cariboea*
23 *Eunice afra*

24 *Marphysa sanguinea*
25 *Eunice filamentosa*
26 unidentified Eunicidae
27 *Lumbrinereis sarsi*
28 *Arabella iricolor*
29 *Eurythoe complanata*
30 unidentified annelids
31 Capitellid sp. 1040
32 *Thelepus setosus*
33 Polycirrus sp. 660
34 Terebellid sp. 837
35 *Nicolea gracilibranchus*
36 unidentified Terebellidae
37 Polydorid sp. 1500
38 *Cirriformia semicincta*
39 *Lygdamis nesiotes*
40 *Sabellastarte indica*
41 unidentified annelids
42 *Thalassema* sp.
43 *Ptychodera flava laysanica*
44 *P. flava laysanica* (tentative)
45 *Octopus* sp.
46 Onuphid

10 B. Menge and Mauzey, 1968 MS *Pisaster* and *Leptasterias* Sink

FOOD WEB MATRIX

	1	2	3	4	5	6
3	1	1	0	0	0	0
4	1	1	0	0	0	0
5	1	1	0	0	0	0
6	1	0	0	0	0	0
7	1	1	1	1	1	1
8	1	1	1	1	1	1
9	1	1	1	1	1	1
10	1	0	0	0	0	0
11	1	1	−1	0	0	1
12	1	1	0	0	0	0
13	1	0	0	0	0	0
14	1	0	0	0	0	0
15	1	0	0	0	0	0
16	1	1	0	0	0	0
17	1	1	0	0	0	0
18	1	1	0	0	0	0
19	1	1	0	0	0	1
20	1	1	0	0	0	1
21	1	0	0	0	0	0
22	1	1	0	0	0	0

KEY

 1 *Leptasterias hexactis*
 2 *Pisaster ochraceus*
 3 *Thais lamellosa*
 4 *Thais canaliculate*
 5 *Thais emarginata*
 6 *Searlesia dira*
 7 *Balanus glandula*
 8 *Balanus cariosus*
 9 *Mytilus edulis*
10 *Chthamalus dalli*
11 *Acmaea scutum*
12 *A. pelta*
13 *A. digitalis*
14 *A. paradigitalis*
15 *Lacuna* spp.
16 *Littorina scutulata*
17 *L. sitchana*
18 *Margarites* spp.
19 *Tonicella lineata*
20 *Katharina tunicata*
21 *Ischnochiton* sp.
22 *Calliostoma ligatum*

11   Niering, 1963   157   Kapingamarangi atoll, Carolines   Community

FOOD WEB MATRIX

	3	4	5	6	7	8	9	11	12	14	15	16	21	22	23	24	25	26	27
1	0	0	1	0	0	0	0	0	0	0	0	0	0	0	0	0	0	0	0
2	1	0	1	0	0	0	0	0	0	0	0	0	0	0	0	0	0	0	0
3	0	1	1	0	0	0	0	0	0	0	0	0	0	0	0	0	0	0	0
4	0	0	1	0	0	0	0	0	0	0	0	0	0	0	0	0	0	0	0
5	0	0	0	1	1	1	1	0	0	0	0	0	0	0	−1	0	0	0	0
10	0	0	0	0	0	0	0	1	0	0	0	0	0	0	0	0	0	0	0
11	0	0	0	0	0	0	1	0	0	0	0	0	0	0	0	0	0	0	0
12	0	0	0	0	0	0	1	0	0	0	0	0	0	0	0	0	0	0	0
13	0	0	0	0	0	0	1	0	1	1	1	0	0	0	0	0	1	0	0
15	0	0	0	0	0	0	1	0	0	0	0	0	0	0	0	0	0	0	0
16	0	0	0	0	0	0	1	0	0	0	0	0	0	0	0	0	0	0	0
17	0	0	0	0	0	0	1	0	0	0	0	1	1	0	0	0	1	1	0
18	0	0	0	0	0	0	1	0	0	0	0	0	0	0	0	0	1	1	0
19	0	0	0	0	0	0	1	0	0	0	0	0	0	0	0	0	1	1	0
20	0	0	0	0	0	0	1	0	0	0	0	0	0	0	0	1	1	1	0
21	0	0	0	0	0	0	0	0	0	0	0	0	1	1	0	0	0	0	0
22	0	0	0	0	0	0	0	0	0	0	0	0	0	1	0	0	0	0	0
27	0	0	0	0	0	0	0	0	0	0	0	0	0	−1	0	0	1	0	0

KEY

1 algae
2 phytoplankton
3 zooplankton
4 invertebrates
5 fish
6 terns
7 frigate birds
8 boobies
9 man
10 turtle grass
11 sea turtles
12 pig
13 coconut
14 rat

15 coconut crabs
16 fowl
17 land vegetation
18 *Cyrtosperma*
19 *Pandanus*
20 breadfruit
21 insects
22 skinks
23 reef heron
24 starlings
25 land crustacea
26 fungi, snails, annelids
27 geckos

FOOD WEB MATRIX

	1	2	3	4	5	6	7	8
2	13	1	0	0	0	0	0	0
3	6	12	0	0	0	0	0	0
4	17	2	0	0	0	0	0	0
5	4	6	0	0	1	0	0	0
6	12	5	1	0	0	0	0	0
7	0	1	0	0	0	0	0	0
8	1	2	0	0	1	0	0	0
9	0	1	0	0	0	0	0	0
10	0	1	0	0	0	0	0	0
11	0	2	1	0	0	0	0	0
12	0	2	8	0	0	0	0	0
13	0	0	0	4	0	0	0	3
14	0	8	13	115	1	81	0	4
15	0	0	0	29	0	0	0	4
16	0	0	2	12	0	5	0	0
17	0	2	0	0	1	0	0	0
18	0	1	0	0	0	0	0	0
19	0	0	0	1	1	0	0	0
20	0	1	0	1	3	0	1	1
21	0	0	2	0	0	1	0	0
22	0	0	2	0	0	1	0	0
23	0	0	4	5	0	0	0	0
24	0	1	1	4	0	0	0	0
25	0	0	0	1	0	0	0	0
26	0	0	0	0	2	0	1	1
27	10	0	0	0	0	0	0	0
28	3	0	0	0	0	0	0	0
29	0	0	16	0	0	0	0	0
30	0	0	15	0	0	0	0	0
31	0	0	0	0	0	0	0	1
32	0	0	1	0	0	0	0	0
33	0	4	10	0	5	1	0	0

KEY

1 *Pleuroploca gigantea*
2 *Fasciolaria tulipa*
3 *F. hunteria*
4 *Busycon contrarium*
5 *B. spiratum*
6 *Murex florifer*
7 *Sinum perspectivum*
8 *Polinices duplicatus*
9 *Conus floridana*
10 *Turbo casteneus*
11 *Urosalpinx* sp.
12 *Nassarius vibex*
13 *Mercenaria campechiensis*
14 *Chione cancellata*
15 *Macrocallista nimbosa*
16 *Cardita floridana*
17 *Aequipecten irradians*

18 *Trachycardium muricatum*
19 *Laevicardium mortoni*
20 *Mactra fragilis*
21 *Anomia simplex*
22 *Ostrea equestris*
23 *Modiolus americanus*
24 *Noetia ponderosa*
25 *Lucina floridana*
26 *Ensis minor*
27 *Atrina rigida*
28 *A. serrata*
29 *Diopatra cuprea*
30 *Onuphis magna*
31 *Owenia fusiformis*
32 *Chthamalus* sp.
33 carrion

APPENDIX 1

13   Richards, 1926   263   Pine feeders, Oxshott Heath   Source

FOOD WEB MATRIX

	2	3	4	5	6	7	8	9	10
1	1	1	0	0	0	0	0	0	0
2	0	0	1	1	0	0	0	0	0
3	0	0	1	0	1	1	1	1	1
4	0	0	0	0	0	0	0	0	1
5	0	0	0	0	0	0	0	0	1
8	0	0	0	0	0	0	0	0	1
9	0	0	0	0	0	0	0	0	1
10	0	0	1	0	0	0	0	0	0

KEY

 1 pine
 2 caterpillars, moths
 3 aphids, secretion
 4 digger wasps
 5 ichneumons
 6 bugs
 7 ants
 8 syrphids
 9 ladybirds
10 spiders

APPENDIX 1

15   Summerhayes and Elton, 1923   232   Bear Island   Community

FOOD WEB MATRIX

	1	2	3	4	5	6	7	8	9	10	11	12	13	14	15	16	17	18	19	20	21	22	24	25	26	27	28
1	0	0	1	1	1	0	0	0	0	0	0	0	0	0	1	0	-1	0	0	0	0	0	0	0	0	0	0
2	0	0	0	0	0	0	0	0	0	0	0	0	0	0	0	0	0	1	0	0	0	0	0	0	0	1	-1
3	0	0	0	0	0	-1	1	1	-1	0	1	1	1	1	0	0	0	0	0	0	0	0	0	0	1	0	0
4	0	0	0	0	0	0	0	0	0	1	0	0	0	0	0	0	0	0	0	0	0	0	0	0	1	1	0
5	0	1	0	0	0	0	0	0	0	0	0	0	0	0	0	0	0	1	0	0	0	0	0	0	1	1	-1
8	0	0	0	0	0	0	0	0	0	0	0	0	0	0	0	0	0	0	0	0	1	0	0	0	0	0	0
9	0	0	0	0	0	0	0	0	0	0	0	0	0	0	0	0	0	0	0	1	0	0	0	0	0	0	0
10	0	0	0	0	0	0	0	0	0	0	0	0	0	0	0	0	0	0	0	0	0	0	0	0	0	1	0
11	0	0	0	0	0	0	0	1	1	0	0	0	0	0	0	0	0	0	0	0	0	0	0	0	0	0	0
12	0	0	0	0	0	0	0	1	1	0	0	0	0	0	0	1	0	0	0	0	0	0	0	0	0	0	0
13	0	0	0	0	0	0	0	1	1	0	0	0	0	0	0	1	0	0	0	0	0	0	0	0	0	0	0
14	0	0	0	0	0	0	0	0	0	0	0	0	0	0	0	1	0	0	0	0	0	0	0	0	0	0	0
15	0	0	0	0	0	0	1	0	0	0	0	0	0	0	0	0	0	0	-2	-2	0	0	0	0	-1	-1	-1
16	0	0	0	0	0	0	0	1	1	0	0	0	0	0	0	0	0	0	0	0	0	0	0	0	0	0	0
18	0	0	0	0	0	0	0	0	0	0	0	0	0	0	0	0	0	0	0	0	0	0	0	0	-1	-1	-1
19	1	0	0	0	0	0	0	0	0	0	0	0	0	0	0	0	0	0	0	0	0	0	0	0	0	0	0
20	1	0	0	0	0	0	0	0	0	0	0	0	0	0	0	0	0	0	0	0	1	0	0	0	0	0	0
21	1	0	0	0	0	0	0	0	0	0	0	0	0	0	0	0	0	1	0	0	1	0	0	0	0	0	0
23	0	0	0	0	0	0	0	0	0	0	0	0	0	0	0	0	0	0	1	1	0	1	0	0	0	0	0
24	0	0	0	0	0	0	0	0	0	0	0	0	0	0	0	0	0	0	0	1	0	1	0	0	0	0	0
25	0	0	0	0	0	0	0	0	0	0	0	0	0	0	0	0	0	0	0	0	0	0	1	0	0	0	0
26	0	0	0	0	0	0	0	1	-1	0	0	0	-1	0	0	1	0	0	1	0	1	0	0	0	0	0	0
27	0	0	0	0	0	0	0	0	0	0	0	0	0	0	0	0	0	1	0	1	0	0	0	0	0	0	0
28	0	0	0	0	0	0	0	0	1	0	0	0	0	0	0	0	0	0	0	0	0	0	0	0	0	0	0

KEY

1 bacteria
2 protozoa (fresh water plankton)
3 plants, dead plants
4 algae (fresh water bottom and littoral)
5 algae (fresh water plankton)
6 worms
7 geese
8 snow bunting, ptarmigan
9 purple sandpiper
10 protozoa (fresh water bottom and littoral)
11 Hymenoptera
12 mites
13 Diptera (adult)
14 Collembola
15 moss
16 spiders
17 protozoa (land)
18 Entomostraca, Rotifera (fresh water plankton)
19 skua, glaucous gull
20 kittiwake, guillemots, fulmar petrel, little auk, puffin
21 northern eider, long-tailed duck, red-throated diver
22 arctic fox
23 marine animals
24 seals
25 polar bear
26 Diptera (immature)
27 Entomostraca, Rotifera, Tardigrada, Oligochaeta, Nematoda (fresh water bottom and littoral)
28 *Lepidurus*

APPENDIX 1

16.1   Paine, 1966   67   Starfish, Mukkaw Bay, Wash.   Sink

FOOD WEB MATRIX

	1	2
2	1	0
3	1	0
4	1	0
5	1	1
6	1	1
7	1	0

KEY

1 *Pisaster*
2 *Thais*
3 chitons
4 limpets
5 bivalves
6 acorn barnacles
7 *Mitella*

16.2   Paine, 1966   68   Starfish, Gulf of California   Sink

FOOD WEB MATRIX

	1	2	3	4	5	6	7
2	1	0	0	0	0	0	0
3	1	0	0	0	0	0	0
4	1	1	0	0	0	0	0
5	1	1	1	0	0	0	0
6	1	1	0	0	0	0	0
7	1	1	0	1	0	0	0
8	1	1	1	0	0	0	0
9	1	1	1	1	0	0	0
10	1	1	1	1	0	0	0
11	1	1	0	0	1	1	1
12	1	0	0	0	0	0	0
13	1	0	0	0	0	0	0

KEY

1 *Heliaster*
2 *Muricanthus*
3 *Hexaplex*
4 *Acanthina tuberculata*
5 *Morula*
6 *Cantharus*
7 *A. angelica*
8 Columbellidae
9 bivalves
10 herbivorous gastropods
11 barnacles
12 chitons
13 brachiopods

APPENDIX 1

16.3   Paine, 1966   69   Gastropods, Costa Rica   Sink

FOOD WEB MATRIX

	1	2	KEY
1	1	0	1 *Thais biserialis*
3	1	0	2 *Acanthina brevidentata*
4	1	1	3 carrion
5	1	1	4 bivalves
			5 barnacles

APPENDIX 1

18   Minshall, 1967   148   Morgan's Creek, Kentucky   Community

FOOD WEB MATRIX

	1	2	3	4	5	6	7	8	9	10	11
6	1	0	0	1	1	0	0	0	0	0	0
7	0	0	1	1	1	1	0	0	0	0	0
8	1	0	1	1	1	1	0	0	0	0	0
9	0	1	1	0	0	1	1	0	0	0	0
10	0	1	1	1	0	0	0	0	0	0	0
11	0	1	1	0	1	1	1	0	0	0	0
12	0	1	0	0	0	1	1	1	1	1	1
13	0	0	0	0	0	0	1	1	1	1	1

KEY

1 Phagocata
2 Decapoda—*Orconectes, Cambarus*
3 Plecoptera—*Isoperia, Isogenus*
4 Megaloptera—*Nigronia, Sialis*
5 Pisces—*Rhinichthys, Semotilus*
6 *Gammarus*
7 Trichoptera—*Diplectrona, Rhyacophila*
8 *Asellus*
9 Ephemeroptera—*Baetis, Centroptilum, Epeorus,
  Paraleptophlebia, Pseudocloeon*
10 Trichoptera—*Neophylax, Glossosoma*
11 Diptera—Tendipedidae, *Simulium*
12 detritus
13 diatoms

153

19   Valiela, 1969   225   Dung   Sink

FOOD WEB MATRIX

	4	5	6	7	8	9
3	1	0	0	0	0	0
4	1	0	0	0	0	0
5	1	0	0	0	1	-1
6	-1	0	0	0	-1	-1
7	1	0	0	0	-1	-1
8	-1	0	0	0	-1	-1
9	0	0	0	0	0	-1
10	1	0	0	0	-1	-1
11	-1	0	0	0	-1	-1
12	0	-1	0	0	0	0
14	1	0	0	0	-1	1
15	1	0	0	0	-1	1
16	1	0	0	0	1	1
18	1	0	0	0	-1	-1
19	0	0	0	0	0	1
21	-1	0	0	0	-1	-1
22	0	0	0	0	0	-1
24	1	0	0	0	1	1
25	1	0	0	0	1	1
26	-1	0	0	0	-1	-1
27	-1	0	0	0	-1	-1
28	-1	0	0	0	-1	-1
30	-1	0	0	0	-1	-1
31	-1	0	0	0	-1	-1
32	-1	0	0	0	-1	-1
33	-1	-1	0	0	-1	-1
34	1	-1	0	0	-1	-1
35	1	-1	0	0	-1	-1
36	1	1	1	1	1	1

KEY

1 *Sphaeridium scarabaeoides*
2 *Sphaeridium scarabaeoides* larva
3 *Cercyon*
4 *Philonthus*
5 *Platystethus*
6 *Falagria*
7 *Oxytelus*
8 *Hyponigrus*
9 *Aleochara*
10 *Atheta*
11 Aleocharine B65
12 *Aerotrichis*
13 *Aphodius*
14 *Musca autumnalis* eggs
15 *Musca autumnalis* first instar
16 *Musca autumnalis* second instar
17 *Musca autumnalis* third instar
18 *Sarcophaga* sp. first instar

19 *Sarcophaga* sp. second instar
20 *Sarcophaga* sp. third instar
21 *Ravinia l'herminieri* first instar
22 *Ravinia l'herminieri* second instar
23 *Ravinia l'herminieri* third instar
24 *Saltella sphondylii* larva
25 *Sepsis* sp. larva
26 Sepsid B larva
27 Leptocera
28 Leptocera larva
29 *Sargus* larva
30 *Psychoda* larva
31 Diptera B66
32 Diptera C66
33 Collembola
34 *Macrocheles*
35 parasitus
36 dung

154

20   Stone, 1969   459   *Chaetognatha*, Agulhas   Sink

FOOD WEB MATRIX

	1	2	3	4	5	6
7	10.1	0.0	0.0	0.0	0.0	0.0
8	0.0	7.5	0.0	0.0	0.0	9.2
9	26.2	22.7	23.7	0.0	10.6	71.3
10	0.0	9.3	0.0	0.0	12.5	0.0
11	0.0	0.0	11.6	0.0	0.0	0.0
12	0.0	29.9	41.5	23.0	24.0	7.1
13	35.7	0.0	0.0	0.0	0.0	0.0
14	1.2	10.4	12.5	47.6	30.9	6.7

KEY

1 *S. serratodentata*
2 *S. enflata*
3 *P. draco*
4 *S. lyra*
5 *S. hexaptera*
6 *S. bipunctata*
7 *Centropages*
8 *Corycaeus*
9 *Corycella*
10 *Eucalanus*
11 *Microsetella*
12 *Oncaea*
13 *Undinula*
14 *other chaetognaths*

23   Reynoldson and Young, 1963   183   Lake-dwelling triclads   Sink

FOOD WEB MATRIX

	1	2	3	4
5	0	10	17	24
6	0	4	8	50
7	0	0	0	35
8	0	0	0	47
9	9	4	3	50
10	6	6	3	45
11	0	1	15	0
12	0	1	11	0

KEY

1 *Polycelis nigra*
2 *P. tenuis*
3 *Dugesia lugubris*
4 *Dendrocoelum lacteum*
5 Oligochaeta—*Lumbriculus variegatus*
6 Arthropoda—*Asellus meridianus*
7 Arthropoda—*Gammarus pulex*
8 Arthropoda—*Cloeon dipterum*
9 Arthropoda—*Diura bicaudata*
10 Arthropoda—*Chironomus* sp. larvae
11 Mollusca—*Hydrobia jenkinsi*
12 Mollusca—*Limnaea pereger* or *Ancylus lacustris*

APPENDIX 1

24   Teal, 1962   616   Salt marsh, Georgia   Community

FOOD WEB MATRIX

	2	3	5	6	7
1	1	0	0	1	0
2	0	1	1	0	0
3	0	0	1	0	0
4	0	0	0	1	0
5	0	0	0	1	0
6	0	0	1	0	1
7	0	0	1	0	0

KEY

1 *Spartina*
2 *Prokelisia, Orchelimum,* other herbivorous insects
3 spiders, passerines, dragonflies
4 algae
5 bacteria
6 *Uca* and *Sesarma, Modiolus, Littorina,* Oligochaete, *Streblospio, Capitella, Manayunkia*
7 *Eurytium,* clapper rail, raccoon

157

25   Harrison, 1962   60   Rain forest, Malaysia   Community

FOOD WEB MATRIX

	2	3	4	5	7	8	9	10	11
1	1	0	1	1	0	0	0	0	0
4	1	1	0	0	1	1	0	1	0
6	0	0	0	0	1	0	0	0	0
7	0	0	0	0	0	-1	0	0	0
9	0	0	1	1	0	0	0	1	1
11	0	0	0	0	0	0	0	1	0

KEY

1 canopy—leaves, fruits, flowers
2 canopy animals—birds, fruit bats and other mammals
3 upper air animals—birds and bats, insectivorous, carnivorous
4 insects
5 large ground animals—large mammals and birds
6 trunk, fruit, flowers
7 middle zone scansorial animals—mammals in both canopy and ground zones
8 middle zone flying animals—birds and insectivorous bats
9 ground—roots, fallen fruit, leaves and trunks
10 small ground animals—birds and small mammals
11 fungi

26   Thomas, 1962   185   River Teify, West Wales   Sink

FOOD WEB MATRIX

	1	2	3
4	1	1	1
5	1	1	1
6	1	1	1
7	1	1	1
8	1	1	0
9	1	0	0
10	1	1	1
11	1	1	1
12	1	1	1
13	1	1	1
14	1	1	1
15	1	1	1
16	1	0	1
17	1	1	1

KEY

1 trout
2 salmon
3 eel
4 Plecoptera
5 Ephemeroptera
6 Trichoptera
7 Megaloptera
8 Hemiptera
9 Odonata
10 adult Coleoptera
11 larval Coleoptera
12 Diptera
13 Mollusca
14 Crustacea
15 Hirudinea
16 Vertebrata
17 terrestrial fauna

27   Hartley, 1948   7   Fish, River Cam, England   Sink

FOOD WEB MATRIX

	1	2	3	4	5	6	7	8	9	10	11
12	0	0	0	0	0	0	0	0	0	0	1
13	0	1	0	0	0	0	0	0	0	0	0
14	0	0	0	0	0	0	0	0	0	0	1
15	1	0	0	0	0	0	0	0	0	0	0
16	3	5	0	0	0	0	0	0	0	0	0
17	7	3	0	0	0	0	0	0	0	0	3
18	2	0	0	0	0	0	0	0	0	0	0
19	3	2	0	0	0	0	0	0	0	0	0
20	3	2	0	0	0	0	0	0	0	0	0
21	20	2	0	0	0	0	0	0	0	0	1
22	0	12	0	0	0	0	0	0	0	0	0
23	0	0	1	0	0	0	0	0	0	2	0
24	0	1	189	0	107	107	0	0	0	1	43
25	0	0	0	0	0	0	0	0	0	0	1
26	0	27	0	0	1	0	0	0	0	4	1
27	0	1	0	0	0	0	0	0	0	0	0
28	0	0	0	0	0	0	0	0	0	0	2
29	0	0	0	0	0	10	0	0	0	0	0
30	0	0	0	1	0	1	0	0	0	0	0
31	0	0	1	3	0	0	0	0	0	1	0
32	0	0	0	3	0	0	0	0	0	0	0
33	1	0	0	0	0	0	1	0	0	0	6
34	1	7	0	1	7	0	0	0	2	0	4
35	0	4	0	0	0	0	0	0	0	0	3
36	1	8	87	2	5	0	1	0	1	22	1
37	1	0	0	0	0	0	0	0	0	1	0
38	0	0	0	0	20	0	0	0	0	0	1
39	0	48	82	3	30	23	4	0	8	11	10
40	0	0	0	0	0	0	0	0	0	0	1
41	0	0	0	1	5	1	0	0	0	3	8
42	0	0	0	0	1	0	0	0	0	1	0
43	0	3	19	0	8	5	1	2	9	1	4
44	0	0	0	0	0	0	0	0	0	0	2
45	0	0	0	0	1	1	0	0	0	1	3
46	0	0	0	0	0	0	0	0	0	0	2
47	0	2	2	0	5	0	0	0	10	0	1
48	0	0	1	0	0	0	0	0	0	0	0
49	0	0	5	0	0	0	0	0	0	0	2
50	0	0	1	0	0	0	0	0	0	0	3
51	0	0	0	0	0	0	0	0	1	0	0
52	0	1	7	0	0	0	3	0	1	0	0
53	0	1	0	0	37	5	0	0	1	0	5
54	0	0	4	0	24	0	0	0	2	0	2
55	0	0	0	0	2	0	0	0	0	0	7
56	0	0	0	0	2	0	0	0	0	0	22
57	0	4	15	3	87	6	2	0	2	1	35
58	0	0	0	0	7	0	0	0	0	0	2
59	0	4	0	0	3	1	0	0	0	0	1
60	0	0	0	0	0	0	0	0	0	0	3
61	0	1	5	0	1	3	0	1	1	0	6
62	0	0	0	0	0	0	0	0	0	0	1
63	0	0	0	0	1	0	0	0	0	0	0
64	0	0	1	0	0	0	0	0	0	0	0
65	0	0	0	0	0	0	0	0	0	0	1
66	0	0	0	0	0	0	0	0	1	0	0
67	0	2	11	3	31	6	0	0	0	0	95
68	0	0	1	0	0	0	0	0	1	0	0
69	0	0	0	0	10	2	0	0	0	1	0
70	0	0	1	3	6	0	0	0	0	0	3

FOOD WEB MATRIX (continued)

	1	2	3	4	5	6	7	8	9	10	11
71	0	0	0	0	0	0	0	0	1	0	0
72	0	0	13	0	0	0	0	0	0	0	0
73	0	1	4	0	0	1	0	0	8	0	0
74	0	37	376	4	93	70	25	0	5	6	3
75	0	0	1	0	0	6	0	0	0	0	0
76	0	0	28	0	0	7	2	0	0	25	0
77	0	0	0	0	0	2	0	0	1	0	0
78	0	0	110	2	0	0	2	0	1	2	0
79	3	4	1565	115	94	55	16	0	68	285	0
80	0	1	23	9	4	5	0	0	2	21	0
81	0	0	0	2	0	0	0	0	0	0	0
82	1	0	1	0	0	0	0	0	0	0	0
83	0	0	0	0	0	0	0	0	1	0	0
84	0	0	7	7	8	8	0	0	0	0	80
85	1	43	0	4	1	0	0	0	3	0	3
86	6	225	281	5	67	34	17	13	40	2	24
87	1	2	0	0	1	0	0	0	8	1	1
88	0	1	0	0	0	0	0	0	0	0	0
89	0	0	135	8	0	0	5	0	0	35	0
90	0	0	0	1	0	0	0	0	0	0	0
91	0	0	145	5	0	8	2	0	1	215	0
92	0	1	53	0	0	0	0	0	0	24	0
93	0	1	0	0	0	0	0	0	0	0	1
94	0	0	3	0	0	0	0	0	0	2	0
95	0	1	0	0	0	0	0	0	3	0	0

KEY

1 pike
2 eel
3 gudgeon
4 minnow
5 dace
6 roach
7 loach
8 perch
9 bullhead
10 3-spined stickleback
11 brown trout
12 *Apodemus sylvaticus*
13 *Microtus* sp.
14 *Rana temporaria*
15 *Anguilla anguilla*
16 *Phoxinus phoxinus*
17 *Gobio gobio*
18 *Rutilus rutilus*
19 *Nemacheilus barbatula*
20 *Cottus gobio*
21 *Gasterosteus aculeatus*
22 *Lampetra fluviatilis*
23 *Sphaerium* sp.
24 *Hydrobia jenkinsi*
25 *Ancylus fluviatilis*
26 *Limnaea pereger* and sp.
27 *Planorbis* sp.
28 Aranaeid indet.
29 Limnocarid indet.
30 Hygrobatid indet.
31 Hydracarinids indet.
32 Myriapoda indet.
33 *Ephemera* sp., adult
34 *Ephemera vulgata* and sp., nymph
35 *Leptophlebia cincta* and sp., nymph
36 *Baetis* sp., nymph
37 *Centroptilum* sp., nymph
38 Ephemeropteran indet., adult
39 Ephemeropteran indet., nymph
40 *Corixa striata*
41 Corixids indet., adult
42 Corixids indet., nymph
43 *Sialis lutaria*, larva
44 *Rhyacophila dorsalis*, larva
45 *Hydroptila* sp., larva
46 *Polycentropus* sp., larva
47 *Hydropsyche* sp., larva
48 *Phryganea* sp., larva

49 *Leptocerus* sp., larva
50 *Molanna angustata* and sp., larva
51 *Glyphotaelius* sp., larva
52 *Limnophilus* sp., larva
53 *Anabolia nervosa*, larva
54 *Stenophylax* sp., larva
55 *Halesus* sp., larva
56 Trichopteran indet., adult
57 Trichopteran indet., larva
58 moths indet., adult
59 caterpillars indet.
60 *Dytiscus marginalis* and sp., adult
61 *D. marginalis* and sp., larva
62 *Hydroporus* sp., adult
63 *Hygrobia hermanni*, adult
64 *Dryops* sp., larva
65 *Gyrinus* sp. adult
66 *Helodes* sp., larva
67 Coleopteran indet., adult
68 Coleopteran indet., larvae
69 Chalcidoids indet., adult
70 Hymenopterans indet., adult
71 *Tipula* sp., larva
72 *Dicranota* sp., larva
73 Tipulid indet., larva
74 *Simulium* sp., larva
75 *Simulium* sp., pupa
76 *Tanypus* sp., larva
77 *Corynoneura* sp., larva
78 *Forcipomyia* sp., larva
79 *Chironomus* sp. and chironomid indet., larva
80 Chironomid indet., pupa
81 Culicid indet., pupa
82 Tabanid indet., larva
83 Anthomyiid indet., larva
84 Dipterans indet., adult
85 *Potamobius pallipes*
86 *Gammarus pulex*
87 *Asellus aquaticus* and sp.
88 *Oniscus* sp.
89 *Chydorus* sp.
90 *Cyclops* sp.
91 Copepoda indet.
92 Ostracoda indet.
93 earthworm indet.
94 Oligochaete indet.
95 *Herpobdella octoculata* and sp.

28.1   Fryer, 1957   217   Rocky shore, Lake Nyasa   Community

FOOD WEB MATRIX

	1	2	3	4	5	6	7	8	9	10	11	12	13	14	15	16	17	18	19	20	21	22	23	24	25	26	30	31
4	1	1	1	0	0	0	0	0	0	0	0	0	0	0	0	0	0	0	0	0	0	0	0	0	0	0	0	0
5	0	1	0	0	0	0	0	1	1	1	0	1	1	0	0	0	0	0	0	0	0	0	0	0	0	0	0	0
15	0	0	1	0	0	0	0	1	1	1	1	1	1	0	0	1	1	1	1	1	1	1	1	1	1	1	1	1
16	0	0	1	0	0	0	1	1	0	1	0	1	1	0	0	1	0	0	0	0	0	0	0	0	1	0	0	0
18	0	0	0	0	0	0	1	1	1	0	0	0	1	0	0	1	1	1	0	0	0	1	0	0	1	0	0	0
19	0	0	0	0	0	0	0	0	1	0	1	0	1	0	0	0	1	0	0	0	0	0	0	0	0	0	0	0
21	0	0	0	0	0	0	0	0	0	0	0	0	1	0	0	0	1	0	1	1	1	1	1	0	1	1	1	0
23	0	0	0	0	0	0	1	1	0	1	1	1	1	1	0	1	1	0	0	0	0	0	0	0	1	0	0	0
24	0	0	0	0	0	0	0	0	1	0	0	0	1	1	0	1	1	0	0	0	0	0	0	0	1	0	0	0
25	0	0	1	0	0	0	0	0	1	1	0	1	1	1	0	1	1	0	1	1	1	1	1	1	1	1	1	0
26	0	0	1	0	1	1	1	1	1	1	1	1	1	1	0	0	1	0	0	0	0	1	0	0	1	1	1	0
27	0	0	1	1	1	1	1	0	0	1	1	1	1	1	0	1	1	1	1	0	1	1	1	1	1	0	1	1
28	0	0	0	0	0	0	0	0	1	0	0	0	0	-1	0	0	1	0	0	0	0	1	0	0	0	0	0	0
29	0	0	1	0	0	0	1	1	0	1	0	0	0	0	0	0	1	0	0	0	1	0	0	0	0	0	1	1

KEY

1 Haplochromis pardalis, other predatory fishes
2 Haplochromis polyodon
3 Haplochromis kiwinge
4 juvenile cichlids
5 Pseudotropheus zebra
6 Pseudotropheus elongatus, P. tropheops, P. minutus, P. auratus, P. fuscus, Labeotropheus fuelleborni, L. trewavasae
7 Haplochromis guentheri, H. fenestratus
8 Mastacembelus shiranus
9 Bathyclarias worthingtoni
10 Labidochromis vellicans
11 Labidochromis caeruleus
12 Haplochromis euchilus
13 Haplochromis ornatus
14 Pseudotropheus fuscoides
15 Hydropsychid larvae
16 Neoperla spio nymphs
17 Cynotilapia afra
18 Potamonautes lirrangensis
19 Schizopera consimilis
20 leech
21 ostracods
22 Barilius microcephalus
23 Eubrianax larvae
24 Elmid larvae
25 Chironomid larvae
26 Afronurus nymphs
27 Aufwuchs
28 terrestrial insects
29 plankton
30 Melanochromis melanopterus
31 Varicorhinus nyasensis

FOOD WEB MATRIX

	1	2	3	4	5	6	7	8	9	10	11	12	13	14	15	16	17	18	19	20	21	22	23	24	25	26	27	28	29	30	36	37	38	39
7	0	0	1	1	1	1	0	0	0	0	0	0	0	0	0	0	0	0	0	0	0	0	0	0	0	0	0	0	0	0	0	0	0	0
8	0	1	1	1	0	1	0	1	0	0	0	0	0	0	0	0	0	0	0	0	0	0	0	0	0	0	0	0	0	0	0	0	0	0
10	0	0	1	0	0	0	0	0	1	1	0	0	0	0	0	0	0	0	0	0	0	0	0	0	0	0	0	0	0	0	0	0	0	0
14	0	0	1	0	0	0	0	0	0	0	0	0	0	0	0	0	0	0	0	0	0	0	0	0	0	0	0	0	0	0	0	0	0	0
20	1	1	1	0	0	0	0	1	0	0	1	0	0	0	1	0	1	0	0	0	0	0	0	0	0	0	0	0	0	0	0	0	0	0
21	1	0	1	0	0	0	0	0	0	0	1	0	0	0	0	0	1	1	0	0	0	0	0	0	0	0	0	0	0	0	0	0	0	0
22	1	0	1	0	0	0	0	0	0	0	1	0	0	0	0	1	1	1	0	0	0	0	0	0	0	0	0	0	0	0	0	0	0	0
23	1	0	1	0	0	0	0	0	0	0	0	0	0	0	0	0	1	1	1	0	0	0	0	0	0	0	0	0	0	0	0	0	0	0
24	0	0	1	0	0	0	0	0	0	0	0	0	0	0	0	0	1	0	0	0	0	0	0	0	0	0	0	0	0	0	0	0	0	0
26	1	0	1	0	0	0	0	0	0	0	0	0	0	0	0	0	0	0	0	0	0	0	0	0	1	1	0	0	0	0	0	0	0	0
27	1	0	1	0	0	0	0	0	0	0	0	0	0	0	0	0	0	0	0	0	0	0	0	1	1	0	1	0	0	0	0	0	0	0
31	1	0	1	0	0	0	0	0	0	0	0	0	0	1	0	0	0	0	0	0	0	0	0	0	1	0	1	0	0	0	1	1	0	1
32	1	0	1	0	0	0	0	0	0	0	0	0	0	0	0	0	0	0	0	0	0	0	0	1	1	0	0	0	0	0	0	0	0	0
33	1	0	1	0	0	0	0	0	0	0	0	0	0	0	0	0	0	0	0	0	0	0	0	1	1	0	0	0	0	0	0	0	0	0
34	0	0	1	0	0	0	0	0	0	0	0	0	0	0	0	0	0	0	0	0	0	0	0	0	0	0	0	0	0	0	0	0	0	0
35	0	0	1	0	0	0	0	0	0	0	0	0	0	0	0	0	0	0	0	0	0	0	0	0	0	0	0	0	0	0	0	0	0	0
37	0	0	1	0	0	0	1	0	0	0	0	0	0	0	0	0	0	0	0	0	0	0	0	0	0	0	0	0	0	0	0	0	0	0
39	1	0	1	0	0	0	0	0	0	0	0	0	0	0	0	0	0	0	0	0	0	0	0	0	0	0	0	0	0	0	0	1	1	0

KEY

1 *Haplochromis johnstoni*
2 *Barbus rhoadesii*
3 *Haplochromis dimidiatus*
4 *Haplochromis rostratus*
5 *Haplochromis* sp.
6 *Rhamphochromis* spp.
7 *Engraulicypris sardella*
8 various immature fishes
9 *Lethrinops brevis*
10 *Tilapia saka-squamipinnis*
11 *Synodontis zambesensis*
12 *Lethrinops*
13 *Alestes imberi*
14 *Microcyclops nyasae*
15 *Lethrinops furcifer*
16 *Barbus eurystomus*
17 *Haplochromis mola*
18 *Barilius microcephalus*
19 *Barbus johnstoni*
20 Baëtid nymphs

21 Ostracods
22 Chironomid larvae
23 *Caridina nilotica*
24 *Corbicula africana*
25 *Barbus innocens, Haplochromis similis, Haplochromis moori*
26 *Lanistes procerus*
27 *Melanoides tuberculata*, other gastropods
28 *Tilapia melanopleura*
29 *Tilapia shirana*
30 *Labeo mesops*
31 terrestrial insects
32 Aufwuchs on *Vallisneria*
33 *Vallisneria*
34 bottom detritus, algae
35 plankton
36 *Haplochromis chrysonotus*
37 *Engraulicypris sardella*
38 *Barbus johnstoni*
39 Cyclopoid copepods

APPENDIX 1

28.3   Fryer, 1957   219   Crocodile Creek, Lake Nyasa   Community

FOOD WEB MATRIX

	1	2	3	4	5	6	7	9	10	11	12	13	14	15	16	17	18	19	20	21	22	23	24	25	26	28	30	32
7	1	0	0	0	0	0	0	0	0	0	0	0	0	0	0	0	0	0	0	0	0	0	0	0	0	0	0	0
8	1	0	0	0	0	0	0	0	0	0	0	0	0	0	0	0	0	0	0	0	0	0	0	0	0	0	0	0
9	0	0	0	0	1	1	0	0	0	0	0	0	0	0	0	0	0	0	0	0	0	0	0	0	0	0	0	0
12	0	0	0	1	0	0	0	0	0	0	0	0	0	0	0	0	0	0	0	0	0	0	0	0	0	0	0	0
13	0	0	1	1	0	0	0	1	0	0	0	0	0	0	0	1	0	0	0	0	0	0	0	0	0	0	0	0
15	0	1	0	0	0	0	0	0	0	0	0	0	0	0	0	0	0	0	0	0	0	0	0	0	0	0	0	0
18	0	0	0	0	1	0	0	0	0	0	0	0	0	0	0	0	0	0	0	0	0	0	0	0	0	0	0	0
19	0	1	1	1	0	0	0	0	0	0	0	1	0	0	0	0	0	0	0	0	0	0	0	0	0	0	0	0
20	0	1	1	1	0	0	0	1	0	0	0	0	0	1	0	0	1	0	0	0	0	0	0	0	0	0	0	0
21	0	0	0	1	0	0	0	0	0	1	0	0	0	0	0	0	0	0	0	0	0	0	0	0	0	0	0	0
22	0	0	0	1	0	0	0	0	0	0	0	0	0	1	0	1	0	0	0	0	0	0	0	0	0	0	0	0
27	0	1	1	0	0	0	1	0	1	0	1	1	0	1	0	1	1	1	1	1	1	1	0	1	1	1	0	0
29	0	0	0	0	0	0	0	0	1	0	0	0	0	0	0	0	0	0	0	0	0	0	0	0	0	1	1	0
31	0	0	0	0	0	0	0	0	0	0	0	0	0	0	0	0	0	0	0	0	0	0	0	0	0	0	1	0
33	0	0	0	0	0	0	0	0	0	1	0	0	0	0	1	0	0	0	0	0	0	0	1	0	0	0	1	1

KEY

1 *Crocodilus niloticus*
2 juvenile Cichlidae
3 *Barbus innocens*
4 *Serranochromis robustus*
5 *Naucoris* sp.
6 *Clarias mellandi*
7 frogs
8 Dytiscid beetles
9 Anisopterid larvae
10 *Barbus paludinosus*
11 *Alestes imberi*
12 mosquito larvae
13 Cyclopoid copepods
14 Zygopterid larvae
15 *Caridina nilotica*
16 *Barilius microcephalus*
17 *Tilapia shirana, T. saka-squamipinnis*

18 *Gyraulus costulatus*
19 Cladocera
20 Chironomid larvae
21 caddis larvae
22 Baëtid nymphs
23 *Micronecta*
24 *Barbus* sp.
25 *Segmentorbis angustus*
26 *Limnaea* sp.
27 bottom algae and detritus
28 *Haplochromis similis*
29 higher plants
30 *Clarias mossambicus*
31 indet. fishes
32 Gerrids
33 terrestrial insects

29   Parsons and LeBrasseur, 1970   341   Strait of Georgia, B.C.   Sink

FOOD WEB MATRIX

	1	2	3	4	5
2	1	0	0	0	0
3	1	0	0	0	0
4	0	0	1	0	0
5	1	0	0	1	0
6	0	1	1	0	1
7	0	0	1	0	1

KEY

1 juvenile pink salmon
2 *P. minutus*
3 *Calanus* and Euphausiid furcilia
4 Euphausiid eggs
5 Euphausiids
6 *Chaetoceros socialis* and *debilis*
7 mu-flagellates

The Mean and Variance of Niche Overlap According to Six Food Web Models

by Thomas Mueller and Joel E. Cohen

Let w be a given food web matrix, with m rows, $n \geq 2$ columns and entries w_{ij} of 0 or 1. Let a, b, c, d refer to distinct columns of w; $a \neq b \neq c \neq d$ throughout. When $n < 4$, expressions involving quartets of distinct columns are 0. When $n < 3$, such triples are 0. If $\Sigma_{i=1}^{m} w_{ia}w_{ib} > 0$, the niches of predators a and b overlap. Let $w_{+j} = \Sigma_{i=1}^{m} w_{ij}$ be the jth column sum, $w_{i+} = \Sigma_{j=1}^{n} w_{ij}$ be the ith row sum, and $A = \Sigma_{i=1}^{m} w_{i+}$ be the matrix sum of w.

Let W be a random $m \times n$ matrix, with elements $W_{ij} = 0$ or 1. The distribution of W is specified by one of the six models. Thus w refers to an observed food web matrix and W to a random matrix; otherwise the conventions for w and W are the same. Define the random variable $B(a,b) = 1$ if $\Sigma_{i=1}^{m} W_{ia}W_{ib} > 0$, $B(a,b) = 0$ otherwise. $P[B(a,b) = 1]$ is the probability, under some model, of niche overlap between predators a and b. The random variable $C = \Sigma_{a<b} B(a,b)$ is the number of edges in the niche overlap graph of W. We seek $E(C) = \Sigma_{a<b} P[B(a,b) = 1] = n(n-1)/2 - \Sigma_{a<b} P[B(a,b) = 0]$, and Var C. In the main text, E is reserved for the actual number of niche overlaps of a real food web w. In this Appendix, E refers only to expectation.

For any random variables X and Y, Var $X = \text{Var}(1 - X)$ and $\text{Cov}(X,Y) = \text{Cov}(1 - X, 1 - Y)$. Hence Var $B(a,b) =$

$P[B(a,b) = 0] - (P[B(a,b) = 0])^2$, $\mathrm{Cov}(B(a,b), B(a,c)) =$
$P[B(a,b) = B(a,c) = 0] - P[B(a,b) = 0] \cdot P[B(a,c) = 0]$ and
$\mathrm{Cov}(B(a,b), B(c,d)) = P[B(a,b) = B(c,d) = 0] - P[B(a,b) = 0]$
$\cdot P[B(c,d) = 0]$. Thus

$$\mathrm{Var}\, C = \Sigma\Sigma(P[B(a,b) = 0] - (P[B(a,b) = 0])^2)$$
$$+ \Sigma\Sigma\Sigma(P[B(a,b) = B(a,c) = 0] - P[B(a,b) = 0] \cdot P[B(a,c) = 0])$$
$$+ \Sigma\Sigma\Sigma\Sigma(P[B(a,b) = B(c,d) = 0] - P[B(a,b) = 0] \cdot P[B(c,d) = 0])$$
$$\tag{2A.1}$$

where the double sum is over the $n(n-1)/2$ pairs (a,b) such that
$a < b$, the triple sum is over the $n(n-1)(n-2)$ triples (a,b,c)
and the quadruple sum is over the $n(n-1)(n-2)(n-3)/4$
quadruples (a,b,c,d) in which $a < b$ and $c < d$.

Model 1: Fixed row sums. Since $P[B(a,b) = 0]$ is the same
for any pair of columns,

$$E(C) = (n(n-1)/2)(1 - P[B(a,b) = 0]). \tag{2A.2}$$

Since rows are mutually independent,

$$\begin{aligned}
P[B(a,b) = 0] &= \Pi_{i=1}^m P[W_{ia} = 0 \text{ or } W_{ib} = 0] \\
&= \Pi_{i=1}^m (1 - P[W_{ia} = W_{ib} = 1]) \\
&= \Pi_{i=1}^m (1 - w_{i+}(w_{i+} - 1)/(n(n-1))).
\end{aligned}$$

Eq. (2A.1) becomes

$$\mathrm{Var}\, C = (n(n-1)/2)(P[B(a,b) = 0] - (P[B(a,b) = 0])^2)$$
$$+ n(n-1)(n-2)(P[B(a,b) = B(a,c) = 0] - (P[B(a,b) = 0])^2)$$
$$+ (n(n-1)(n-2)(n-3)/4)(P[B(a,b) = B(c,d) = 0]$$
$$- (P[B(a,b) = 0])^2). \tag{2A.3}$$

It remains to find $P[B(a,b) = B(a,c) = 0]$ and $P[B(a,b) = B(c,d) = 0]$. Now $P[B(a,b) = B(a,c) = 0]$ is the product, for
$i = 1$ to m, of $P[W_{ia} = 0 \text{ or } (W_{ia} = 1 \text{ and } W_{ib} = W_{ic} = 0)] =$
$(n - w_{i+})/n + w_{i+}(n - w_{i+})(n - w_{i+} - 1)/(n(n-1)(n-2))$.
Similarly $P[B(a,b) = B(c,d) = 0]$ is the product, for $i = 1$ to m, of
$P[(W_{ia} = 0 \text{ or } W_{ib} = 0) \text{ and } (W_{ic} = 0 \text{ or } W_{id} = 0)]$

167

$$= P[W_{ia} + W_{ib} + W_{ic} + W_{id} = 0]$$
$$+ P[W_{ia} + W_{ib} + W_{ic} + W_{id} = 1]$$
$$+ (4/6)P[W_{ia} + W_{ib} + W_{ic} + W_{id} = 2]$$
$$= [n(n-1)(n-2)(n-3)]^{-1} \cdot (\Pi_{k=0}^{3}(n - w_{i+} - k)$$
$$+ 4w_{i+}(n - w_{i+})(n - w_{i+} - 1)(n - w_{i+} - 2)$$
$$+ (4/6)\binom{4}{2}w_{i+}(w_{i+} - 1)(n - w_{i+})(n - w_{i+} - 1)).$$

Model 2: Truncated binomial row sums. Equations (2A.2) and (2A.3) apply again since $P[B(a,b) = 0]$ is the same for all (a,b). Let p_i be the p-parameter of the 0-truncated binomial distribution with mean w_{i+}, obtained as described in section 4.2.1; $q_i = 1 - p_i$. To find $E(C)$, we need only $P[B(a,b) = 0]$ $= \Pi_{i=1}^{m}(1 - P[W_{ia} = W_{ib} = 1]) = \Pi_{i=1}^{m}(1 - p_i^2/(1 - q_i^n))$, excluding from the product those rows i such that $p_i = 0$.

To find Var (C) from equation (2A.3), we also need $P[B(a,b) = B(a,c) = 0]$ and $P[B(a,b) = B(c,d) = 0]$. As in model 1, the first probability is $\Pi_{i=1}^{m}(q_i(1 - q_i^{n-1}) + p_iq_i^2)/(1 - q_i^n)$, excluding i where $p_i = 0$. Again using the intermediate results of model 1, the second probability is $\Pi_{i=1}^{m}[(q_i^4(1 - q_i^{n-4}) + 4p_iq_i^3 + 4p_i^2q_i^2)/(1 - q_i^n)]$, excluding i where $p_i = 0$.

Model 3: Fixed column sums. There are $\binom{m}{w_{+b}}$ possible ways of assigning the w_{+b} 1's in column b to rows. Of these assignments, $\binom{m - w_{+a}}{w_{+b}}$ will result in no overlap with the w_{+a} 1's in column a when $w_{+a} + w_{+b} \le m$, and 0 otherwise. Hence when $w_{+a} + w_{+b} \le m$, $P[B(a,b) = 0] = \binom{m - w_{+a}}{w_{+b}} / \binom{m}{w_{+b}}$, and $= 0$ otherwise. This determines $E(C)$.

Since

$$P[B(a,b) = B(a,c) = 0]$$
$$= \binom{m - w_{+a}}{w_{+b}}\binom{m - w_{+a}}{w_{+c}} / \left[\binom{m}{w_{+b}}\binom{m}{w_{+c}}\right]$$
$$= P[B(a,b) = 0]P[B(a,c) = 0],$$

the triple sums in equation (2A.1) are 0. Since distinct columns are independent, the quadruple sums in equation (2A.1) are 0. Thus $P[B(a,b) = 0]$ determines Var C also.

Model 4: Truncated binomial column sums. Let p_j now be the p-parameter of the 0-truncated binomial distribution with mean w_{+j}, and $q_j = 1 - p_j$.

To find $E(C)$, we calculate $P[B(a,b) = 1] = \Sigma_{j=1}^m P[W_{+a} = j$ and at least one 1 in column b occurs in a row where column a also has a 1] $= \Sigma_{j=1}^m \binom{m}{j} p_a^j q_a^{m-j} (1 - q_b^j)/[(1 - q_a^m)(1 - q_b^m)]$.

With the help of the binomial theorem, this may be transformed to an expression that avoids the possibility of division by 0:
$P[B(a,b) = 1] = \Sigma_{k=0}^{m-1} (q_a + q_b - q_a q_b)^k/[(\Sigma_{k=0}^{m-1} q_a^k)(\Sigma_{k=0}^{m-1} q_b^k)]$.
Alternatively, an expression for $P[B(a,b) = 0]$ follows from noticing that $P[B(a,b) = 0 \mid W_{+a} = j, W_{+b} = k] = \binom{m-j}{k}\Big/\binom{m}{k}$
as in model 3, so that $P[B(a,b) = 0] =$

$$\Sigma_{j=1}^m \Sigma_{k=1}^{m-j} \binom{m-j}{k}\binom{m}{j} p_a^j q_a^{m-j} p_b^k q_b^{m-k}/[(1 - q_a^m)(1 - q_b^m)].$$

Using this and the first formula for $P[B(a,b) = 1]$, it is readily checked that $P[B(a,b) = 0] + P[B(a,b) = 1] = 1$.

Since columns are again independent, the quadruple sum in equation (2A.1) again vanishes. Thus Var (C) is determined by $P[B(a,b) = 1]$ above and by $P[B(a,b) = B(a,c) = 1]$
$= \Sigma_{j=1}^m \binom{m}{j} p_a^j q_a^{m-j} (1 - q_b^j)(1 - q_c^j)/[(1 - q_a^m)(1 - q_b^m)(1 - q_c^m)]$.
In computation, division by 0 can be avoided by canceling a common factor of $p_a p_b p_c$ from numerator and deonominator.

The tidy factorization, in model 3, of $P[B(a,b) = B(a,c) = 0]$, which permits the triple sums to vanish in equation (2A.1), does not hold here.

Model 5: Fixed matrix sum. Again $P[B(a,b) = 0]$ is the same for all pairs of columns, so equations (2A.2) and (2A.3) apply. $P[B(a,b) = 0] = \Sigma_{j=0}^m P[W_{+a} = j$ and no 1 in col-

umn b occurs in a row where column a also has a 1] $= \Sigma_{j=0}^{m}$
$\binom{m}{j}\binom{mn-m-j}{A-j} \Big/ \binom{mn}{A}$. Alternatively $P[B(a,b) = 0] =$
$\Sigma_{j=0}^{m} P[W_{+a} + W_{+b} = j$ and $W_{ia} + W_{ib} \le 1, \ i = 1, \ldots, m] =$
$\Sigma_{j=0}^{m} 2^j \binom{m}{j}\binom{mn-2m}{A-j} \Big/ \binom{mn}{A}$. The non-obvious identity of
these two formulas for $P[B(a,b) = 0]$ was proved by J. Riordan
and R. Donaghey using the known identities $\Sigma_{j=k}^{m} \binom{m}{j}\binom{j}{k} =$
$\binom{m}{k} 2^{m-k}$ and $\binom{N+j}{A+j} = \Sigma_{k=0}^{j} \binom{j}{k}\binom{N}{A+k}$.

$P[B(a,b) = B(a,c) = 0] = \Sigma_{j=0}^{m} P[W_{+a} = j$ and there is a 0
in the rows of columns b and c where there is a 1 in column a]
$= \Sigma_{j=0}^{m} \binom{m}{j}\binom{mn-m-2j}{A-j} \Big/ \binom{mn}{A}$. Finally $P[B(a,b) = B(c,d)$
$= 0] = P[$where there is a 1 in column a, there is a 0 in column
b; and where there is a 1 in column c, there is a 0 in column d]
$= \Sigma_{j=0}^{2m} \binom{2m}{j}\binom{mn-2m-j}{A-j} \Big/ \binom{mn}{A}$.

Model 6: Binomial matrix sum. Since $P[B(a,b) = 0]$ is
the same for any pair of columns, equations (2A.2) and (2A.3)
apply. Moreover, distinct pairs of columns are independent,
so the last term of (2A.3) vanishes. Define $p = A/(mn)$ so that
A is the mean of a binomial distribution; $q = 1 - p$. Then
$P[B(a,b) = 0] = \Pi_{i=1}^{m}(1 - P[W_{ia} = W_{ib} = 1]) = (1 - p^2)^m$.
Explicitly $E(C) = (n(n-1)/2)(1 - (1 - p^2)^m)$. $P[B(a,b) =$
$B(a,c) = 0] = (P[W_{1a} = 0$ or $(W_{1a} = 1$ and $W_{1b} = W_{1c} = 0)])^m$
$= (q + pq^2)^m = (1 - 2p^2 + p^3)^m$. Hence Var $C =$
$n(n-1)[(1 - p^2)^m/2 + (n-2)(1 - 2p^2 + p^3)^m - (n - 3/2)$
$\cdot (1 - p^2)^{2m}]$.

We now prove that the expected number $E_1(C)$ of niche
overlaps under model 1 (fixed row sums) is less than or equal to
the expectation $E_2(C)$ under model 2 (truncated binomial row
sums). It suffices to show $P_2[B(a,b) = 0] \le P_1[B(a,b) = 0]$.
The subscripts refer to the model number.

By a convex function f of a real argument x we mean a real function such that for $x < y$ and $0 \le p \le 1$, $f(px + (1 - p)y) \le pf(x) + (1 - p)f(y)$. A function f is concave if $-f$ is convex. In particular $f(x) = x(x - 1)$ is convex since $f''(x) > 0$. So for fixed n, $1 - x(x - 1)/(n(n - 1))$ is concave. Therefore, for $1 \le x_i \le n$, $\Pi_{i=1}^m (1 - x_i(x_i - 1)/(n(n - 1)))$ is a concave function of each x_i when the remaining x_j, $j \ne i$, are fixed. Therefore by Jensen's inequality (Parzen, 1960, p. 434) and the independence of rows,

$$
\begin{aligned}
P_2[B(a,b) = 0] \\
= E(\Pi_{i=1}^m (1 - x_i(x_i - 1)/(n(n - 1))) \,|\, W_{i+} = x_i) \\
\le \Pi_{i=1}^m (1 - w_{i+}(w_{i+} - 1)/(n(n - 1))) \\
= P_1[B(a,b) = 0]
\end{aligned}
$$

where the expectation is with respect to the independent 0-truncated binomial distributions of W_{i+}, $i = 1, \ldots, m$, which satisfy $E(W_{i+}) = w_{i+}$.

$E_1(C) < E_2(C)$ strictly unless $\max_i w_{i+} = 1$, when $E_1(C) = E_2(C) = 0$, or unless $\max_i w_{i+} = n$, when $E_1(C) = E_2(C) = n(n - 1)/2$.

The same line of argument also proves that $E_4(C) \le E_3(C)$. Since $\binom{n + 1}{k} - \binom{n}{k} = \binom{n}{k - 1}$, which increases with n, where n and $k - 1$ are non-negative integers, $\binom{n}{k}$ is a convex function of n for k fixed. Therefore, for positive integers x and y such that $2 \le x + y \le m$, $\binom{m - x}{y} / \binom{m}{y} = \binom{m - y}{x} / \binom{m}{x}$ is a convex function of x for each y and of y for each x. So by Jensen's inequality and the independence of columns,

$$
P_3[B(a,b) = 0] =
$$

$$
\binom{m - w_{+a}}{w_{+b}} / \binom{m}{w_{+b}} \le E\left[\binom{m - x}{y} / \binom{m}{y} \,\Big|\, W_{+a} = x, W_{+b} = y \right]
$$

$$
= P_4[B(a,b) = 0],
$$

where the expectation is with respect to the independent 0-truncated binomial distributions of $W_{+j}, j = 1, \ldots, n$, which satisfy $E(W_{+j}) = w_{+j}$.

For each of the remaining possible inequalities among the means of these six models, there are counterexamples. Counterexamples can be constructed for all of the possible inequalities among the variances of the six models. Some of these counterexamples are matrices of 0's and 1's which could not be among our food web matrices since they have at least one row or column with no 1's.

Expected Niche Overlaps Assuming Randomly Overlapping One-Dimensional Niches

Let X_1 and X_2 be two independent random variables uniformly distributed on $[0, 1]$. Let $X_{(1)} = \min(X_1, X_2)$, $X_{(2)} = \max(X_1, X_2)$. According to MacArthur's (1957) model of overlapping niches (his Hypothesis II), the niche of a given kind of organism A is defined as the interval $(X_{(1)}^A, X_{(2)}^A)$, where the superscript identifies the organism. The niches of different kinds of organisms are assumed independent.

Ignoring MacArthur's derivations (1957, 1960; Vandermeer and MacArthur, 1966), we calculate the expected number of niche overlaps in a community of n predators. The niches of A and B do not overlap if $X_{(2)}^B < X_{(1)}^A$ or $X_{(2)}^A < X_{(1)}^B$; otherwise they overlap. Since the probability of niche overlap is the same for any pair A and B of kinds of predators, the expected number $E(C)$ of niche overlaps according to this model is $E(C) = \binom{n}{2}(1 - P[\text{niches of } A \text{ and } B \text{ do not overlap}])$. By symmetry, $P[\text{niches of } A \text{ and } B \text{ do not overlap}] = 2P[X_{(2)}^B < X_{(1)}^A]$. The probability density function of $X_{(1)}$ is $g_1(x) = 2(1 - x)$ and the cumulative distribution function of $X_{(2)}$ is $G_2(x) = x^2$, for x in $[0, 1]$ (Kendall and Stuart, 1969, p. 325). By independence of niches, $P[X_{(2)}^B < X_{(1)}^A] = \int_0^1 g_1(x) G_2(x)\, dx = 1/6$. Hence $E(C) = (2/3)\binom{n}{2} = n(n - 1)/3$.

References

Fame is a food that dead men eat,—
I have no stomach for such meat.
                    Henry Austin Dobson

Fame is a bread that feeds the dead
(The living, too, I've heard it said).

The numbers and letters following each reference show where
it is cited. P = Preface; numbers refer to sections of the text;
A = Appendix following text. Fn = Fig. n; Tn = Table n.

Allee, W. C., Alfred E. Emerson, Orlando Park, Thomas Park,
    and Karl P. Schmidt, 1949. *Principles of Animal Ecology*.
    Philadelphia and London: W. B. Saunders. 7.1.1

Altmann, Stuart A., and Jeanne Altmann, 1970. *Baboon Ecology*.
    Basel: S. Karger. 7.2.2

Altschul, Aaron M., 1967. Food protein: new sources from
    seeds. *Science* 158:221–230. 8

Bartholomew, D. J., 1967. *Stochastic Models for Social Processes*.
    London: John Wiley and Sons. 4.2.1

Benzer, S., 1959. On the topology of the genetic fine structure.
    Proc. Nat. Acad. Sci. U.S.A. 45:1607–1620. 2.1

Bird, Ralph D., 1930. Biotic communities of the aspen parkland
    of central Canada. *Ecology* 11:356–442. 2.1, 3.2, T1, T4,
    F1, A1

Birkeland, Charles, 1974. Interactions between a sea pen and
    seven of its predators. *Ecol. Monogr.* 44:211–232. 3.1

Bityukov, see Naumov, 1972, p. 98. F26

Booth, Kellogg S., 1975. *PQ*-tree algorithms. Ph.D. thesis,
    Lawrence Livermore Laboratory, University of California,
    Livermore. 2.1

Booth, Kellogg S., and George S. Lueker, 1976. Testing for the consecutive ones property, interval graphs, and graph planarity using *PQ*-tree algorithms. *J. Comput. Syst. Sci.* 13(3):335–379. 2.1, 7.1.2

Carlson, Clarence A., 1968. Summer bottom fauna of the Mississippi River, above Dam 19, Keokuk, Iowa. *Ecology* 49(1):162–169. 3.1

Carroll, Lewis, 1960, in Martin Gardner (ed.), *Alice's Adventures in Wonderland. The Annotated Alice.* New York: Clarkson N. Potter. 2

Chapman, Royal N., 1931. *Animal Ecology.* New York and London: McGraw Hill. 7.1.1, F25

Clapham, W. B., Jr., 1973. *Natural Ecosystems.* New York: Macmillan. 3.1

Clarke, Thomas A., Arthur O. Flechsig, and Richard W. Grigg, 1967. Ecological studies during Project Sea Lab II. *Science* 157(3795):1381–1389. 3.2, T4, A1

Clements, F. E., and V. E. Shelford, 1939. *Bio-ecology.* New York: John Wiley and Sons. 2.4

Cody, Martin L., 1968. On the methods of resource division in grassland bird communities. *Am. Nat.* 102(924):107–147. 2.3, 6.1.3, 6.2

Cohen, Joel E., 1966. *A Model of Simple Competition.* Cambridge, Mass.: Harvard University Press. 6.2

———, 1977a. Food webs and the dimensionality of trophic niche space. *Proc. Nat. Acad. Sci.* 74(10):4533–36. P.

———, 1977b. Ratio of prey to predators in community food webs. *Nature* 270:165–167. 4.2.2

Daley, Richard, 1977. The man who made Chicago work [Obituary of Richard Daley, former Mayor of Chicago]. *Time Magazine*, 3 January 1977, p. 46. 7.1

Danzer, Ludwig, and Branko Grünbaum, 1968. Intersection properties of boxes in R^d. *Technical Report No. 4* (mimeo), Department of Mathematics, University of Washington, Seattle, Sept. 1968. 7.1.1

Danzer, Ludwig, Branko Grünbaum, and Victor Klee, 1963.

Helly's theorem and its relatives. *Proc. Symp. Pure Math.*, Vol. VII (*Convexity*), 101–180. 7.1.1

Dayton, Paul K., 1973. Two cases of resource partitioning in an intertidal community: making the right prediction for the wrong reason. *Am. Nat.* 197(957):662–670. 3.2

Dendy, J. S., 1945. Predicting depth distribution in three TVA storage type reservoirs. *Trans. Am. Fish. Soc.* 75:65–71. F22

Denison, W. C., 1968. Personal communication. 2.1, F4

Dewdney, A. K., 1977. Embedding graphs in Euclidean 3-space. *Am. Math. Monthly* 84(5):472–373. 7.1.1

Ehrlich, G., S. Even, and R. E. Tarjan, 1976. Intersection graphs of curves in the plane. *J. Combinat. Theory* 21(1): 8–20. 7.1.1

Elton, Charles S., 1966. *The Pattern of Animal Communities.* London: Methuen. 1

Erdös, P., and A. Rényi, 1960. On the evolution of random graphs. *Publ. Math. Inst. Hung. Acad. Sci.* 5:17–61. 5.2, 8.1

Fabré, J. H., 1913. see Elton, 1966, p. 38. 1

Fisher, Sir Ronald A., 1970. *Statistical Methods for Research Workers*, 14th ed. Edinburgh: Oliver and Boyd. 4.2.1

Fryer, Geoffrey, 1959. The trophic interrelationships and ecology of some littoral communities of Lake Nyasa with especial reference to the fishes, and a discussion of the evolution of a group of rock-frequenting Cichlidae. *Proc. London Zool. Soc.* 132:153–281. 3.2, T4, A1

Fulkerson, D. R., and O. A. Gross, 1965. Incidence matrices and interval graphs. *Pac. J. Math.* 15(3):835–855. P, 2.1, 5.3

Gabai, Hyman (in press). *n*-dimensional interval graphs. 7.1.1

Gallopin, Gilberto C., 1972. Structural properties of food webs. In Bernard C. Patten (ed.), *Systems Analysis and Simulation in Ecology*, Vol. 2, 241–282. New York: Academic Press. 1, 2.1, 4.2.3

Gates, David M., 1969. Climate and stability. *Diversity and*

Stability in Ecological Systems, Brookhaven Symposia in Biology, No. 22, 115–127. Upton, New York: Brookhaven National Laboratory. 7.1.1

Gavril, Fănică, 1974. The intersection graphs of subtrees in trees are exactly the chordal graphs. *J. Combinat. Theory (B)* 16:47–56. 7.1.2

Gilman, Leonard, and Allen J. Rose, 1974. *APL: An Interactive Approach.* New York: John Wiley and Sons. P, 5.3

Gilmore, P. C., and A. J. Hoffman, 1964. A characterization of comparability graphs and of interval graphs. *Can. J. Math.* 16:539–548. 2.1

Gilpin, Michael, 1977. Personal communication. 6.1.3

Haigh, J., and J. Maynard Smith, 1972. Can there be more predators than prey? *Theor. Popul. Biol.* 3(3):290–299. 4.2.2

Hairston, Nelson G., 1949. The local distribution and ecology of the Plethodontid salamanders of the Southern Appalachians. *Ecol. Monogr.* 19:47–73. 3.2, 6.1.2, 6.3, T4, F21, A1

Harary, Frank, and Edgar M. Palmer, 1973. A survey of graphical enumeration problems. In J. N. Srivastava et al. (eds.), *A Survey of Combinatorial Theory.* New York: North Holland. 5.1

Hardy, A. C., 1924. The herring in relation to its animate environment. Part 1. The food and feeding habits of the herring with special reference to the East Coast of England. *Fish. Invest. Ser. II* 7(3):1–45. 2.1, 3.1, 3.2, T4, A1

Harrison, J. L., 1962. The distribution of feeding habits among animals in a tropical rain forest. *J. Anim. Ecol.* 31(1):53–63. 3.2, T4, A1

Hartley, P. H. T., 1948. Food and feeding relationships in a community of fresh-water fishes. *J. Anim. Ecol.* 17(1):1–14. 2.4, 3.2, T4, A1

Hiatt, Robert W., and Donald W. Strasburg, 1960. Ecological relationships of the fish fauna on coral reefs of the Marshall Islands. *Ecol. Monogr.* 30(1):65–126. 3.1

Hubert, Lawrence, 1974. Problems of seriation using a subject by item response matrix. *Psychol. Bull.* 81(12):976–983. 2.1

Hutchinson, G. E., 1944. Limnological studies in Connecticut. VII. A critical examination of the supposed relationship between phytoplankton periodicity and chemical changes in lake waters. *Ecology* 25:3–26. 1

————, 1959. Homage to Santa Rosalia, or why are there so many kinds of animals? *Am. Nat.* 93:145–159. 4.2.3, 4.4

————, 1965. *The Ecological Theater and the Evolutionary Play.* New Haven and London: Yale University Press. 1

Jeffries, Clark, Victor Klee, and Pauline van den Driessche (in press). When is a matrix sign stable? *Can. J. Math.* 6.3

Johnson, Norman L., and Samuel Kotz, 1969. *Discrete Distributions.* Boston: Houghton Mifflin. 4.2.1

Kashkarov, D., and V. Kurbatov, 1930. Preliminary ecological survey of the vertebrate fauna in the Central Kara-Kum Desert in West Turkestan. *Ecology:* 11(1): 35–60. 3.1

Kendall, David G., 1969. Incidence matrices, interval graphs and seriation in archaeology. *Pac. J. Math.* 28(3):565–570. 2.1, 7.1.2

————, 1970. A mathematical approach to seriation. *Phil. Trans. Roy. Soc. Lond. A.* 269:125–135. 7.1.2

Kendall, Maurice G., and Alan Stuart, 1969. *The Advanced Theory of Statistics,* Vol. 1., 3rd ed. London: Griffin. A3

Klee, Victor, 1969. What are the intersection graphs of arcs in a circle? *Am. Math. Mon.* 76:810–813. P, 7.1.1

————, 1976. Personal communication. 7.1.1, 7.1.2

Koepcke, H.-W. and M., 1952. Sobre el proceso de transformacion de la materia organica en las playas arenosas marinas del Peru. *Publ. Univ. Nac. Mayor San Marcos, Zoologie* Serie A. No. 8. 3.1, 3.2, 4.2.3, T4, A1

Kohn, Alan J., 1959. The ecology of *Conus* in Hawaii. *Ecol. Monogr.* 29:47–90. 2.1, 2.2, 2.3, 3.2, 5.1, 6.1.2, 6.1.3, 6.4, T4, A1

———, 1976. Personal communication. 3.2

Komlós, János, 1976. Personal communication. 8.1

Krebs, Charles J., 1972. *Ecology: The Experimental Analysis of Distribution and Abundance*. New York: Harper and Row. 7.1.1

Lekkerkerker, C. G., and J. C. Boland, 1962. Representation of a finite graph by a set of intervals on the real line. *Fund. Math. Polska Akad. Nauk.* 51:45–64. 2.1, 6.3

Levin, Simon A. (ed.), 1975. *Ecosystem: Analysis and Prediction.* Proceedings of a SIAM-SIMS Conference on Ecosystems, Alta, Utah, July 1–5, 1974. Philadelphia: Society for Industrial and Applied Mathematics. 7.2.3

Lewontin, Richard C., 1969. The meaning of stability. *Diversity and Stability in Ecological Systems.* Brookhaven Symposia in Biology No. 22, 13–24. Upton, New York: Brookhaven National Laboratory. 7.2.4

Lindeman, R. L., 1942. The trophic-dynamic aspect of ecology. *Ecology* 23:399–418. 3.1, 6.3

MacArthur, Robert H., 1957. On the relative abundance of bird species. *Proc. Nat. Acad. Sci.* 43:293–295. 6.2, 6.4, A3, F23

———, 1960. On the relative abundance of species. *Am. Nat.* 94:25–36. 6.2, 6.4, A3, F23

———, 1972. *Geographical Ecology: Patterns in the Distribution of Species.* New York: Harper and Row. 2.4, 6.1.3, 7

Maguire, Bassett, Jr., 1973. Niche response structure and the analytical potentials of its relationship to the habitat. *Am. Nat.* 107(954):213–246. 7.1.1

Mauzey, Karl P., 1966. Feeding behavior and reproductive cycles in *Pisaster ochraceus*. *Biol. Bull.* 131:127–144. 3.2

———, 1967. The interrelationship of feeding, reproduction, and habitat variability in *Pisaster ochraceus*. Ph.D. thesis, Department of Zoology, University of Washington, Seattle. 3.2, T4, A1

May, Robert M., 1973, *Stability and Complexity in Model Ecosys-*

tems. Princeton: Princeton University Press. 6.1.3, 6.2, 6.3, 6.4, 8.1

———, 1974. On the theory of niche overlap. *Theor. Popul. Biol.* 5(3):297–332. 6.2, 6.3, 6.4, 8.1

———, 1975. Some notes on estimating the competition matrix α. *Ecology* 56:737–741. 2.4, 7.2.1

———, 1977. Personal communication. 7.2.3

Menge, Bruce A., 1970. The population ecology and community role of the predaceous asteroid, *Leptasterias hexactis* (Stimpson). Ph.D. thesis, Department of Zoology, University of Washington, Seattle. 3.2, T4, A1

———, 1972. Competition for food between two intertidal starfish species and its effect on body size and feeding. *Ecology* 53(4):635–644. 3.2

Miller, Richard S., 1967. Pattern and process in competition. In J. B. Cragg (ed.), *Adv. Ecol. Res.* 4:1–74. London: Academic. 1, 2.4

Miller, Rupert G., 1974. The jackknife—A review. *Biometrika* 61:1–15. 7.2.2

Minshall, G. Wayne, 1967. Role of allochthonous detritus in the trophic structure of a woodland springbrook community. *Ecology:* 48(1):139–149. 3.1, 3.2, T4, A1

Mosteller, C. F., and J. W. Tukey, 1968. Data analysis, including statistics. In G. Lindzey and E. Aronson (eds.), *Revised Handbook of Social Psychology*. Reading, Mass.: Addison-Wesley. 7.2.2

Nash, Leonard K., 1963. *The Nature of the Natural Sciences.* Boston: Little, Brown. 7.2.1

Naumov, N. P., 1972. *The Ecology of Animals*, Norman D. Levine (ed.), Frederick K. Plous, Jr. (translator). Urbana: University of Illinois Press. 7.2.1, 7.2.4

Niering, William A., 1963. Terrestrial ecology of Kapinga-marangi Atoll, Caroline Islands. *Ecol. Monogr.* 33:131–160, Spring. 3.2, T4, A1

Odum, Eugene P., 1971. *Fundamentals of Ecology*, 3rd ed. Philadelphia: W. B. Saunders. F22

Olson, Everett C., 1966. Community evolution and the origin of mammals. *Ecology* 47(2):291–302. 3.1

Paine, Robert T., 1963. Trophic relationships of eight sympatric predatory gastropods. *Ecology* 44(1):63–73. 3.2, 6.1.2, T4, A1

———, 1966. Food web complexity and species diversity. *Am. Nat.* 100(910):65–75. 3.2, T4, A1

———, 1977. Personal communication. 7.2.2

Parker, see Chapman, 1931, p. 110. F25

Parsons, T. R., and R. J. LeBrasseur, 1970. The availability of food to different trophic levels in the marine food chain. In J. H. Steele (ed.), *Marine Food Chains*, 325–343. Berkeley and Los Angeles: University of California Press. 3.2, T4, A1

Patten, Bernard C. (ed.), 1975. *Systems Analysis and Simulation in Ecology*, Vol. 3. New York: Academic Press. 7.2.3

Parzen, Emanuel, 1960. *Modern Probability Theory and Its Applications*. New York: John Wiley and Sons. A2

Peirce, Charles S., 1872. Educational textbooks [Book review of J. C. Maxwell's *Theory of Heat*]. *The Nation* 14:244–246. Reprinted in K. L. Ketner and J. E. Cook (eds.), *Contributions to The Nation 1869–1893*. Lubbock, Texas: Texas Tech Press. 1975, pp. 46–51. P

Petipa, T. S., E. V. Pavlova, and G. N. Mironov, 1970. The food web structure, utilization and transport of energy by trophic levels in the planktonic communities. In J. H. Steele (ed.), *Marine Food Chains*, 143–167. Edinburgh: Oliver and Boyd. 3.1

Pianka, Eric R., 1974. *Evolutionary Ecology*. New York: Harper and Row. 3.1

———, 1976. Competition and niche theory. In R. M. May (ed.), *Theoretical Ecology: Principles and Applications* 114–

181

141. Oxford: Blackwell Scientific. 1, 2.4, 6.2.3

Pulliam, H. Ronald, 1969. Personal communication. 6.2

Reynoldson, T. B., and J. O. Young, 1963. The food of four species of lake-dwelling triclads. *J. Anim. Ecol.* 32(2):175–191. 3.2, T4, A1

Richards, O. W., 1926. Studies on the ecology of English heaths. III. Animal communities of the felling and burn successions at Oxshott Heath, Surrey. *J. Ecol.* 14:244–281. 3.2, T4, A1

Roberts, Fred S., 1969a. On the boxicity and cubicity of a graph. In W. T. Tutte (ed.), *Recent Progress in Combinatorics*, 301–310. New York: Academic Press. 7.1.1

———, 1969b. On nontransitive indifference. RM-5782-PR. Santa Monica: The Rand Corporation. 32 pp. 2.1

———, (ed.), 1976a. *Energy: Mathematics and Models*. Proceedings of a SIMS Conference on Energy held at Alta, Utah, July 7–11, 1975. Philadelphia: Society for Industrial and Applied Mathematics. 7.2.3

———, 1976b. Structural analysis of energy systems. In Fred S. Roberts (ed.), *Energy: Mathematics and Models*, 84–101. Proceedings of a SIMS Conference on Energy held at Alta, Utah, July 7–11, 1975. Philadelphia: Society for Industrial and Applied Mathematics. 7.2.3

———, 1976c. *Discrete Mathematical Models*. Englewood Cliffs, N.J.: Prentice-Hall, Inc. P

——— (in press). Food webs, competition graphs, and the boxicity of ecological phase space. In Y. Alavi and D. Lick (eds.), *Theory and Applications of Graphs—In America's Bicentennial Year*. New York: Springer Verlag. P

Root, Richard B., 1975. Some consequences of ecosystem texture. In Simon A. Levin (ed.), *Ecosystem: Analysis and Prediction*, 88–97. Proceedings of a SIAM-SIMS Conference on Ecosystems, Alta, Utah, July 1–5, 1974. Philadelphia: Society for Industrial and Applied Mathematics. 404 pp. 2.1, 2.2, 3.1

Russell, Clifford S. (ed.), 1975. *Ecological Modeling: In a Resource Management Framework*. The Proceedings of a Symposium Sponsored by National Oceanic and Atmospheric Administration and Resources for the Future. Washington: Resources for the Future, Inc. 7.2.3

Schoener, T. W., 1967. The ecological significance of sexual dimorphism in size in the lizard *Anolis conspersus*. *Science* 155:474–477. 6.1.1, F20

———, 1971. Theory of feeding strategies. *Annu. Rev. Ecol. Syst.* 2:369–404. 8.1

———, 1974. Resource partitioning in ecological communities. *Science* 185:27–39. 2.3, 6.1.1, 6.1.3, 6.3, 6.4, 7.2.1

———, 1977. Personal communication. 2.1, 2.2, 7.2.3

Shelford, Victor E., 1913. Animal communities in temperate America as illustrated in the Chicago region: A study in animal ecology. *Geographic Society of Chicago Bulletin 5*. Chicago: University of Chicago Press. 1, 3.1

Sissman, L. E., 1967. String Song. *Dying: An Introduction*, 68–70. Boston: Little Brown. 5

Snedecor, G. W., and W. G. Cochran, 1967. *Statistical Methods*, 6th ed. Ames, Iowa: The Iowa State University Press. 4.2.1, 5.3

Spitzer, Frank, 1964. *Principles of Random Walk*. Princeton: D. Van Nostrand. 6.3

Stevens, Wallace, 1971. *Collected Poems*, p. 78. New York: Knopf. 3

Stone, James H., 1969. The Chaetognatha community of the Agulhas Current: Its structure and related properties. *Ecol. Monogr.* 39(4):433–463, Autumn. 3.1, 3.2, T4, A1

Summerhayes, V. S., and C. S. Elton, 1923. Contributions to the ecology of Spitsbergen and Bear Island. *J. Ecol.* 11:214–286. 3.1, 3.2, T4, A1

Teal, John M., 1962. Energy flow in the salt marsh ecosystem of Georgia. *Ecology* 43(4):614–624. 3.2, T4, A1

Thomas, J. D., 1962. The food and growth of brown trout

(*Salmo trutta* L.) and its feeding relationships with the salmon parr (*Salmo salar* L.) and the eel (*Anguilla anguilla* L.) in the River Teify, West Wales. *J. Anim. Ecol.* 31(2): 175–205. 3.2. T4, A1

Valiela, Ivan, 1969. An experimental study of the mortality factors of larval *Musca autumnalis* DeGeer. *Ecol. Monogr.* 39(2):199–225, Spring. 3.2, T4, A1

Vandermeer, John H., 1969. The competitive structure of communities: An experimental approach with Protozoa. *Ecology* 50(3):362–371. 8.2

———, 1972. Niche theory. *Annu. Rev. Ecol. Syst.* 3:107–132. 1

——— and R. H. MacArthur, 1966. A reformulation of alternative (b) of the broken stick model of species abundance. *Ecology* 47(1):139–140. A3

Voltaire, 1770. *Letter to d'Alembert.* 6

Watt, Kenneth E. F., 1975. Critique and comparison of biome ecosystem modeling. In Bernard C. Patten (ed.), *Systems Analysis and Simulation in Ecology*, Vol. 3, 139–152. New York: Academic Press. 7.2.3, 7.2.4

Whittaker, R. H., 1969. Evolution of diversity in plant communities. *Diversity and Stability in Ecological Systems*. Brookhaven Symposia in Biology, No. 22, 178–195. Upton, New York: Brookhaven National Laboratory. 2.1

Wiegert, Richard G. (ed.), 1976. *Ecological Energetics*. Benchmark Papers in Ecology, Vol. 4. Stroudsburg, Pa.: Dowden, Hutchinson and Ross. 7.2.3

Zaret, Thomas M., and R. T. Paine, 1973. Species introduction in a tropical lake. *Science* 182:449–455. 8.2

Index

185

INDEX

dominant clique defined, 15; dominant clique matrix, 15, 116; defined, 16
Donaghey, R., 170
Drosophila, 99
dung, 29, 38, 154

eating relation, 13, 21, 23, 121
eats, defined, 118
ecology, community, 109; physiological, 109
economics, 19
edge, 58, 81
eel, 39
Eltonian food-chain, 58; predator chain, 57
energetics, 99
energy, 99, 122; flow, 36
England, 160
enumeration of interval graphs, 75
error, 86; of observation, 20
evolution, Darwinian, 124
expectation, 166
expected frequencies of interval graphs, 82

fame, 174
feeding relation, 13, 21, 23, 121
fish, 29, 160; predatory, 39
Florida, 147; sand bar, 37
food chain defined, 56; maximal, 59; defined, 56
food web defined, 3, 14; formalized, 9; graph, 12, 18; matrix, 14, 166; food web matrix defined, 13; model, 58, 166; seasonal, 118; time-dependence of, 130
forest, 95

game fish, 100–101
gastropods, 29, 32, 37, 97, 147, 152
gecko, 36
genetic fine structure, 19
genetics, Mendelian, 124
Georgia, 38, 157
grain beetles, 113
graph, 9; defined, 12
grasshopper eggs, 113
grassland bird communities, 98, 100
guild, 7
Gulf of California, 151

gulls, 34

habitat, 98, 104; defined, 21; factors, 6
Hawaii, 32–33, 98, 122, 142–44
Heliaster, 37
Helly's theorem, 115
herbivore, 11, 57
herring, 29, 36, 139
Hillel, 41
home range, 121
host, 133
humidity, 105, 113–14

incidence matrix defined, 14
independence of niche dimensions, 93, 98, 112–13, 115
independent sampling, 78
index, 185
inequalities, 61, 128, 170–71; Jensen's, 171
insectivores, 16
insect trapping, 121
interpolation, 77, 84
intersection graph, 12, 112, 114, 116; defined, 111
interval, 7; representation, 11; interval food web defined, 20; interval graph, 16, 19, 41, 115, 127; applications of, 19; defined, 12; interval pseudo-random graphs, 82

jackknife, 121, 130
Jensen's inequality, 171

Kapingamarangi atoll, 29, 36, 146
Kentucky, 37, 153
kinds of organism, 7
kittiwakes, 34

Lake Nyasa, 29, 35, 59, 162–64
leaf litter, 37
lemon, 75
length of chain, 59; of maximal food chains, 127; of predatory lizard, 93; of prey, 93; of snail, 97; of shell, 97
Leptasterias, 29, 36, 145
life cycles, 18
lime, 75
littoral, 35
lizard size, 94

186

INDEX

Washington, 150
water loss, 113
weight of food, 100
Whittaker, Anne, 1–190
willow, 31; communities, 8, 12; forest, 10, 14–15, 134

worms, 41

yield/effort curve, 120, 121, 130

zero-truncated binomial distribution, 58–60, 168–71

LIBRARY OF CONGRESS CATALOGING IN PUBLICATION DATA

Cohen, Joel E
    Food webs and niche space.

    (Monographs in population biology; 11)
    Bibliography: p.
    Includes index.
    1. Food chains (Ecology) 2. Niche (Ecology)
I. Title. II. Series.
QH541.C62   574.5′3   77-85534
ISBN 0-691-08201-4
ISBN 0-691-08202-2 pbk.